Environmental Crime and Corruption in Russia

Environmental devastation, a significant consequence of industrial activity in Soviet times, continues to be a major problem in Russia. Specific problems include radioactive pollutants from inadequately monitored nuclear plants; illegal logging and wildlife poaching, which have grown into hugely profitable businesses for criminal gangs; and toxic waste from unsanctioned and poorly controlled metallurgical, petroleum and agricultural chemical industries. This book presents a wide-ranging assessment of the environmental problems faced by Russia, and of the crime and corruption that contribute to them. It concludes, gloomily, that the problems are getting worse and that little is being done to tackle them.

Sally Stoecker is a scholar-in-residence, School of International Studies, American University, Washington, DC, USA. She received her doctorate from the Paul H. Nitze School of Advanced International Studies, Johns Hopkins University, Washington, DC, USA.

Ramziya Shakirova is a research analyst at Independent Project Analysis, Inc. and holds a doctorate in Public Policy from the School of Public Policy, George Mason University, Arlington, VA, USA.

Routledge Transnational Crime and Corruption Series

Published in association with the Terrorism, Transnational Crime and Corruption Center, School of Public Policy, George Mason University, USA

Environmental Crime and Corruption in Russia

Federal and regional perspectives

Edited by Sally Stoecker and Ramziya Shakirova

Routledge
Taylor & Francis Group

LONDON AND NEW YORK

First published 2014 by Routledge

2 Park Square, Milton Park, Abingdon, Oxfordshire OX14 4RN
711 Third Avenue, New York, NY 10017

Routledge is an imprint of the Taylor & Francis Group, an informa business

First issued in paperback 2017

British Library Cataloguing in Publication Data
A catalogue record for this book is available from the British Library

Library of Congress Cataloging in Publication Data
Environmental crime and corruption in Russia : federal and regional perspectives / edited by Sally Stoecker and Ramziya Shakirova.
    pages ; cm. – (Routledge transnational crime and corruption series ; 8)
    Includes bibliographical references and index.
    1. Offenses against the environment – Russia (Federation)
    2. Environmental protection – Corrupt practices – Russia (Federation)
    3. Environmental law – Russia (Federation) – Criminal provisions.
    4. Corruption – Russia (Federation) 5. Russia (Federation) –
    Environmental conditions. I. Stoecker, Sally W., editor.
    II. Shakirova, Ramziya, editor. III. Series: Routledge transnational crime
    and corruption series ; 8.
    HV6405.R8E58 2013
    364.1´450947–dc23   2013010921

ISBN: 978-0-415-69870-2 (hbk)
ISBN: 978-0-8153-7467-1 (pbk)

Typeset in Times New Roman
by HWA Text and Data Management, London

Contents

Illustrations

Figures

Tables

Maps

Contributors

Elena Bocharnikova graduated from the Soil Science Department, Moscow State University, in 1984. In 1991 Bocharnikova obtained her doctoral degree. The title of her thesis was "The Influence of Oil Pollution on Properties of Gray-Brown Soil of Apsheron and Gray Forest Soil of Bashkiria." More recently, she conducted field research in Siberia, Azerbaijan, and Central Russia, where industrial soil pollution occurs. Since 1994, she has been employed by the Institute of Physical-Chemical and Biological Problems in Soil Science at the Russian Academy of Sciences. Bocharnikova is a specialist in the area of environmental pollution from heavy metals, hydrocarbons and phosphorus, and, in particular, the biogeochemical cycle of silicon in the soil-plant-microbial system, including migration, transformation and interaction of soluble silicon compounds with heavy metals and hydrocarbons. She has published widely and has conducted research in the United States, both at Florida State University and at California State University. She also has patents for her green technologies that protect the environment.

Nailya Davletova is a candidate of medical sciences at the Kazan State Medical University of the Ministry of Health and Social Welfare, Tatarstan, Russia and is employed as an assistant in the Department of Hygiene, Occupational Medicine and Medical Ecology. Her scholarly interests include hygiene, the ecology of man, and social ecology. She has written three books and more than 50 articles. As a student, she became interested in environmental pollution. She began to wonder why the health of the population of Tatarstan was deteriorating and how to rectify the situation. When she was awarded the V. I. Vernadsky scholarship, her fate was sealed. As it had with hundreds of other talented young people, the scholarship brought Davletova to the scientific heritage of academician V. I. Vernadsky. It drew her attention to issues of environmental protection and balanced economic development. The research that she conducted in graduate school revealed that in the Republic of Tatarstan there is much damage from pollution. Currently, she is implementing her findings to protect the environment and improve the ecological situation in Tatarstan.

Alexander Nakhabov graduated from the Obninsk Institute for Nuclear Power Engineering in 2002, specializing in the instrumentation and methods of quality control and diagnostics. In 2009 he received his doctorate in technical sciences.

Currently, Nakhabov is Associate Professor and Deputy Head of the Automatics, Testing and Diagnostics department at the Obninsk Institute. He lectures on control theory, physical methods of nondestructive testing, and nondestructive testing of nuclear power plants. He has published numerous papers and reports. In fall 2009 he visited the United States through the "Open World" Nonproliferation Program. His research interests include: analysis of nondestructive testing and diagnostics of nuclear power plants (NPP) based on data from mining methods, and the development of computer-aided systems for processing and analyzing nondestructive testing and diagnostics of NPP.

Larisa Pervushina is a scientist with the Udmurt Association of Waste Recyclers, a non-governmental organization created in 2006. Among the main goals of the association is to create an information base, make professional contacts, and provide legal support of their interests in state environmental agencies. In a short time, the association has become a trusted member of Udmurtia's professional society. One of the largest projects they have participated in is focused on recycling and disposal of waste and also drafting of the law "On industrial waste and demand in the Udmurt Republic."

Yekaterina Pisareva graduated from the N. G. Chernyshevsky Trans-Baikal State Humanitarian-Pedagogical University in 2010 with a thesis entitled "Criminal Responsibility for the Illegal Felling of Forest Plantations in Russia." Currently, she is working on her doctoral dissertation titled, "Methods of Investigating Forest Fire Crimes," focusing on issues related to criminal responsibility for corruption and other white-collar crimes in this area. Her research is vital to environmental protection because crimes such as bribery, malversation, criminal negligence, forgery by officials and others pervade crimes leading to forest fires. Pisareva was a recipient of TraCCC's Saratov Center's research grant in 2011 to conduct a research on methods of combating corrupt practices and crime in the forest sector of Trans-Baikal region.

Dmitry Samokhin is a senior instructor in the Reactor Development and Design Department of the Obninsk Institute for Nuclear Power Engineering in Moscow. Among the courses he teaches are "The physical theory of nuclear reactors" and "The dynamics and safety of nuclear power plants." Samokhin has published numerous reports and articles, including contributions to the *Encyclopedia of Nuclear Power Plant Violations*. In Spring 2009, he visited the United States through the "Open World" Nonproliferation Program. His research interests include the development of methods and safety of reactor plants, and the study and design of nuclear reactors and nuclear power plants.

Ramziya Shakirova is a research analyst at Independent Project Analysis, Inc. and holds a doctorate in Public Policy from the School of Public Policy, George Mason University, Arlington, VA, USA. She holds a candidate of economic sciences degree from the Kazan Institute of Economics and Finance, and taught graduate and undergraduate courses at the Kazan Institute of Finance and Economy in Tatarstan.

She is the author of *State Property Management in Transitional Economies* (1998) and co-author of the textbook *The Economy of Tatarstan* (2001).

Sally Stoecker is a scholar-in-residence, School of International Studies, American University, Washington, DC, USA. She received her doctorate in Eurasian studies and international relations from Johns Hopkins University, Paul H. Nitze School of of Advanced International Studies (SAIS), in 1995. For 13 years she was a research associate with the RAND Corporation, analyzing Soviet military and foreign policy. Since 1998 she has been researching various aspects of transnational crime and has spent considerable time in Siberia and the Russian Far East. Currently she is focusing on environmental crimes in Russia and recently managed the research and publication of two training manuals on investigating illegal logging and poaching in the Russian Far East.

Svetlana Tulaeva is a doctoral student at the University of Eastern Finland in the Department of Social Science and Business Studies. Her dissertation topic is the regulation of nature management resources in Russia. She is researching new ways for companies to improve their legality, focusing on the conflicting requirements of different regulatory systems (Russian legislation, intergovernmental conventions, international non-government environmental standards promoted by NGOs, local rules, internal corporate governance). She became interested in environmental problems while employed by the Centre for Independent Social Research in St. Petersburg. Together with her colleagues from the Centre she took part in a study of the Russian forestry business and the Forest Stewardship Council. Her attendance at seminars and summer school sessions at the Institute for the Rule of Law at the European University at St. Petersburg deepened her knowledge and understanding of law and corruption.

Acknowledgments

The authors would like to thank, first of all, Peter Sowden, commissioning editor for Routledge books on Asian, Middle Eastern and Post-Soviet Studies, for supporting our project on environmental crimes in Russia. Few books containing research by Russian scholars are available in English and we hope this volume will spur interest and further research in this area.

Karen Saunders and Louise Shelley of the Terrorism, Transnational Crime and Corruption Center (TraCCC) at George Mason University helped turn the idea of a book into a reality by encouraging us to craft a proposal for submission to Routledge. TraCCC has a series of books on transnational crime with Routledge but none on environmental crime. It seemed like an important and logical addition. Louise Shelley also provided constructive comments on portions of the book.

Many TraCCC staff, friends, and colleagues were collectively indispensable to the completion of the book. Aaron Beitman and Karen Saunders performed first-rate translations of some chapters from Russian to English; Caitlin Kurylo formatted graphs and tables and handled any technical issue with ease; Tim Dempsey translated bibliographies. A few chapters were very technical in nature and this behooved us to find specialists to read through our translations for accuracy and readability. Carol Christensen, PhD epidemiologist at EPA, read and commented on Nailya Davletova's study of the health impact of pollution; Vorneen Hultman, an environmental engineer, reviewed Elena Bocharnikova's chapter on construction materials made with industrial waste.

Last but not least we want to thank our beloved editor, Joyce Horn, for her rigor, competence, and remarkable patience. Whether or not she ever thought she would edit so many books and papers on Russian crime is unclear, but she has been a loyal TraCCC editor and great friend for many years.

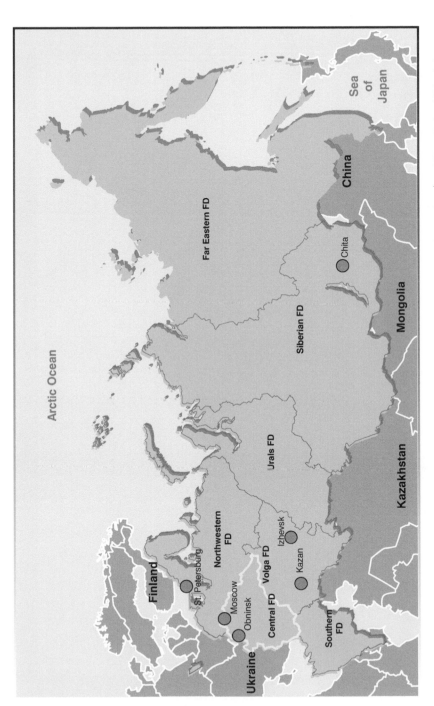

Map 1.1 The locations where the authors conducted research are shown on the map of the Russian Federation (FD = Federal District)

Introduction

The sanctity of the earth's land, water and air is under severe challenge from many manmade phenomena, including global warming, the spread of disease and contaminants across borders, the despoiling of clean water supplies, the destruction of rainforests, and the demand for a higher standard of living from the growing world population. An important and related issue is environmental crime: the illegal harvesting of valuable forest resources; unregulated industrial pollution; the poaching of wildlife; and the illegal construction of roads and structures that destroy natural habitats (flora and fauna). Environmental crime frequently involves toxic and radioactive materials, and can directly harm human health. The latter is of particular concern in Russia given its "peacetime demographic crisis"[1] resulting from, among other factors, human exposure to contaminated air, soil, and water. Russian environmentalists have frequently accused the government of violating their rights to a clean and healthy environment, as provided for in the Russian constitution, and of not compensating victims of environmental crimes and harm.[2]

The consequences of environmental crimes are quite serious: low life expectancy; contamination of food and water, which undermines health; lack of proper control of disposable wastes; organized crime operating with a natural resource model; and disposal of resources without thought to the future and their replacement.

Unfortunately, there is a dearth of Russian research (and translations into English) on these subjects and, so inhibiting our capacity to understand how crimes are perpetrated in Russia and what legal, administrative, and technical tools are available to address environmental offenses and protect natural resources.

This volume on environmental degradation and crime in Russia presents a collection of papers by Russian scholars detailing acts which negatively affect Russia's unique and vast forests, natural resources, fresh water supplies, and human health. In addition, corruption exacerbates pollution and contributes to the plundering of natural resources sold by criminal entrepreneurs to the highest bidder. Thus, we explore environmental crimes and corruption, and their interdependence.

One of the biggest challenges confronting Russia, as well as many other Warsaw Pact and NATO countries today is how to dispose of the radioactive waste accumulated during the cold war from huge arsenals of nuclear weapons. Aleksei Yablokov,[3] President Boris Yeltsin's advisor on environmental affairs in the 1990s, meticulously researched and catalogued environmental threats to Russia stemming from nuclear accidents during the cold war. Inadequate and even harmful efforts to contain and dispose of nuclear and chemical wastes that accumulated are surveyed,[4] such as the Soviet navy dumping tons of radioactive waste into the oceans surrounding the USSR, nuclear-powered submarines lost at sea or decaying in port, and grave health impacts from the high levels of air and water pollution from the largest nickel and palladium plant in the world, Norisk Nikel.[5] In 2004, Yablokov estimated the total economic losses from environmental degradation in Russia was the equivalent of 10–20 percent of GNP.

Although Russia has access to the most water resources in the world with 30.2 thousand cubic meters per person/per year, a large percentage of the water cannot be consumed because it is contaminated with industrial and agro-industrial waste.[6] Outdated and/or poorly functioning technology cannot adequately purify the wastewater from the housing, industrial, and agricultural economies. The destruction of cold-war era missiles has contaminated the water quality in Udmurtia and other regions.[7]

Some of Russia's environmental crimes today occur due to negligence and corruption of government officials, while others are a result of premeditated actions by organized crime groups or corrupt individuals. Industrial pollution often results from profit-driven enterprises employing outdated technologies whose owners ignore environmental requirements and regulations. These regulations, in turn, are rarely enforced by officials who are often corrupt. Many industrial enterprises still employ decades-old technologies and equipment incapable of recycling waste or capturing toxic emissions. Though there is considerable discussion of implementing green technology in many industrial plants, it does not appear to have taken root. The global economy and the presence of economic powers around Russia, particularly China, provide a large market for valuable natural resources. Illegal markets have emerged to sell Russia's valuable natural resources for private gain. Illicit Chinese wood processors make deals with Russian loggers known as "black lumberjacks" and import tons of wood from Russia illegally. Presently, Russia does not have sufficient capacity, especially in the Far East, to process the wood and sell finished products at home and abroad. The illegal felling and clandestine shipping of high quality timber to China can be called, if not capital flight, "cut and run."[8] This not only harms the Russian environment, but also the state coffers. No taxes are paid on valuable wood that is smuggled out of the country by criminals and sold in the Chinese and other Asian markets.

The destruction of unique forest resources in Russia that absorb a tremendous amount of atmospheric carbon dioxide, as a result of illegal logging, is also jeopardizing the health of the entire planet as well as the habitats that support rare forms of flora and fauna.[9]

This book is unique because of the lack of such research available in English. The authors were identified by the Terrorism, Transnational Crime and Corruption

Center (TraCCC) of George Mason University in Arlington, Virginia through a grant program in 2010–11 sponsored by the US Department of State. When the scholars completed their individual research for TraCCC, it became evident that an impressive amount of current research on environmental crime and corruption affecting many Russian regions from St. Petersburg in the northwest to Chita in eastern Siberia. Through a mutual process of critique and review between the TraCCC translators and the authors, this series of reports on Russian environmental topics was completed, translated and is now accessible to an English-reading audience in this volume.

The editors of the book would like underscore that environmental policy is constantly evolving and changes have occurred since the chapters were written and authors' assertions and recommendations may be irrelevant or outdated. For example, during the final editing of the book in July 2013, President Putin signed into law stronger punishments for poachers of endangered wildlife. For the first time in Russian legal history, irrespective of the cost and condition of the wildlife contraband uncovered – whether in parts, derivatives, or the entire animal – the perpetrator will face criminal charges.

On January 19, 2010, the Southern Federal District was split in two, with its former southern territories forming a new North Caucasian Federal District and the total number of FDs rising to eight. Most of the research here was completed before that change and therefore data on the North Caucasian FD is not included.

These Russian scholars, interdisciplinary in training and diverse in regional geographic spread, examine criminal environmental phenomena occurring on their soil. Not all environmental crimes in Russia are covered in this volume nor do we provide guidance for investigating them. Rather, we hope to inspire future research into these issues which affect not only Russia but the health of the planet. Finally, we wish to thank all the authors for allowing us to publish their work in English, so a broader audience can benefit from it.

Chapter 1: Envisaging environmental crime in Russia: past and present realities

The co-editors of the volume, Sally Stoecker and Ramziya Shakirova, former researchers at TraCCC, provide an overview of environmental crime in Russia. They review motivations for corrupt and criminal behavior, the legacies of the past when a command-driven economy forced industries to fulfill ambitious plans irrespective of the cost to the environment, the effectiveness of laws and technologies, and today's profit-driven ambitions that have an equally devastating effect on the environment that provide habitats for multitudes of flora and fauna. They also explore and compare the pollution indicators across federal districts and attempt to illustrate where public health problems are most concentrated and why.

Chapter 2: Russia's nuclear industry and the environment

Dmitry Samokhin and Alexander Nakhabov, scientists and instructors with the Reactor Development & Design Department of the Obninsk Institute for Nuclear Power Engineering near Moscow, write about many aspects of the nuclear industry and its impact on the environment. They analyze the power structures and the international and domestic laws that attempt to secure nuclear materials and keep them out of the hands of criminals. Massive construction projects currently under way in Russia to produce more energy contribute to the large amount of nuclear waste already in existence. Disposing of that waste is a daunting task that requires considerable skill and expertise, and Samokhin and Nakhabov discuss ways in which it can be improved.

Chapter 3: Forest auctions in Russia: how anti-corruption laws facilitate the development of corrupt practices

Svetlana Tulaeva, a research associate with the Center for Independent Social Research in St. Petersburg, examines the scourge of corruption in the Russian forest sector, where a set of informal and corrupt norms compensate for the absence of an effective market economy. According to the Russian Forest Code, leases of forests should take place through open forest auctions—a form of competitive bidding. In theory, auctions ensure fair competition for all participants and discourage informal relationships between officials and businessmen. In reality, however, they lead to collusion between the buyers and new corrupt practices. Most common is a conspiracy between an official representative of the seller (a civil servant) and the prospective buyer, as well as a conspiracy among prospective buyers. As a result, these practices lead to the under-pricing of forest leases and a decline in state revenues. Market competition among the participants in state auctions is replaced by informal interactions. In essence, the anti-corruption law expunges corruption at one level and produces new corrupt practices at another.

Chapter 4: Combating corruption and organized crime in the forest sector of the Trans-Baikal Territory

Yekaterina Pisareva is a doctoral student at the N.G. Chernyshevsky Transbaikal State Humanitarian and Pedagogical University in Chita, Russia. She wrote her dissertation on methodologies for investigating forest fire crimes. In the chapter included here, she analyzes the role of organized crime in illegal logging in Chita and districts of Eastern Siberia based on responses to a sociological survey of law enforcement officers, investigators, and prosecutors at police and forest agencies responsible for addressing forest crimes. Illegal logging harms national and regional economies, she finds. Profits that the state should have earned from the sale of timber often end up in the pockets of private individuals and various organizations dealing with the forestry business. The uncontrolled development of market relations, the undervaluing of and lack of control over natural resources, and inadequate legislation, have resulted in numerous violations of the law. The situation is further complicated by the fact that China—the largest importer of timber and timber products—is in close proximity of the Trans-Baikal region and their demands for Siberian wood have grown exponentially. Former President Medvedev issued a decree "On the National Anti-Corruption Strategy and the National Anti-Corruption Plan for 2010–2011" to eradicate the causes and conditions leading to corruption in Russian society, but time will tell if this has any impact on the present-day problem of illegal logging.

Chapter 5: Environmental crimes on the territory adjacent to the petroleum-storage facility in the town of Kama in the Kambarsk Region of the Udmurt Republic

Larisa Pervushina is a research associate with the Association of Waste Recycling in Izhevsk, Udmurtiia, a non-governmental organization, created in 2006. The success of her NGO is evident in the programs they developed and the laws they helped sponsor to ensure the recycling and proper burial of industrial waste. Her chapter addresses the pollution of petroleum products as environmental crimes and the threats to environmental security on the property of the petroleum enterprise in the Kama Kambarskii region of the Udmurt Republic. The territory surrounding the enterprise has been severely damaged from pollution, thereby violating article 42 of the Constitution that guarantees every citizen the right to a clean environment. Oil extraction, production, processing, transportation, storage and disposal are all activities to which effective cleansing and protective measures must be applied. Currently, these processes, from extraction to production, are poorly regulated and pose hazards to operators. The author investigates the environmental degradation surrounding enterprises and documents accidents that have occurred, permitting oil to leak into the soil, and the lack of methods for evaluating the cost to the environment. She reviews regulations and makes recommendations for improving them.

Chapter 6: The impact of metallurgical and cement industrial waste on Central Russia's environment

Elena Bocharnikova, senior scientist at the Institute of Physical-Chemical and Biological Soil Studies, Russian Academy of Sciences, Moscow, examines the heavy metals and radionuclides contained in metal, cement and chemical waste of industrial and agro-industrial enterprises. She illustrates how enterprises have illegally used these wastes as soil ameliorant and construction material for paving roads. Moreover, data about the content of contaminants in industrial waste is often down-played, underestimated or intentionally "reduced" so that the enterprises can obtain certificates allowing them to use this waste in agriculture and construction. She finds that although laws and regulations were developed in the Soviet era for the containment and disposal of toxic waste, today they are simply ignored. This in turn leads to tons of waste entering the atmosphere, water and soil.

Chapter 7: Environmental contamination and public health in the Republics of Tatarstan and Mari El, Russian Federation

Nailya Davletova, Candidate of Medical Sciences, Kazan, Tatarstan, analyzes medical and epidemiological data to convincingly illustrate that contaminated drinking water and air pollution have a deleterious impact on public health in the republics of Tatarstan and Mari El. In order to grasp the scale and scope of the pollution, a comprehensive, multi-faceted investigation of the environment and its impact on health was undertaken. Her research reveals that drinking water, soil and the atmosphere contribute most significantly to what the Russian scientists refer to as "environmental-hygienic degradation" or "environmental indicator of poor health" (hereafter abbreviated as EIOPH, the Russian acronym). Twelve districts in the Republic of Tatarstan, three districts and one city in Mari El, were found to have poor drinking water. As for air pollution, twelve districts and the city of Kazan in Tatarstan, and one municipal district and three cities in Mari El were adversely affected. Illustrating the pollution problems with graphs and tables, she urges preventative measures to be taken to warn citizens of these health hazards and efforts to clean up the air and water to begin.

Notes

1 N. Eberstadt, *Russia's Peacetime Demographic Crisis: Dimensions, Causes, Implications,* Seattle, WA: National Bureau of Asian Research, 2010.
2 A.V. Yablokov, *Zelenoe dvizhenie i grazhdanskoe obshchestvo: narushenie ekologicheskikh prav grazhdan v Rossii,* Moscow: KMK, 2004, 10–11.
3 Dr. Alexei Yablokov is a Corresponding Member (Biology) of the Russian Academy of Sciences, and Honorary Member of the American Academy of Arts and Sciences. Among many other positions, he served as Counselor for Ecology and Public Health to the Russian President in the early 1990s.
4 V.I. Bulatov, "Remediation of polluted areas in the Ob-Irtysh Basin," Cristian Ion, ed. *Second National Dialogue: Forum on Atomic Energy, Society, and Security,* 21–22 April 2008, St. Petersburg," sponsored by Green Cross Russia, Green Cross Switzerland, and Global Green USA. Speaking about the nuclear accident at the Chemical Complex in Seversk (near Tomsk) in 1993, Bulatov stated that the quantities of radioactive waste are so great that 50 reservoirs are needed to contain it.
5 L. Bronder, A. Nikitin, V. Nikiforov, and I. K. Utgiver, "Environmental challenges in the Arctic – Norilsk Nickel: The Soviet legacy of industrial pollution," Bellona Foundation, 2010, http://www.bellona.no/rapporter/norilsk-nickel-report-en (accessed 22 March 2012).
6 A.L. Terian, "The implementation of state water control: improving legal regulations," in V.R. Koniakin and M.L Prokhorova, eds., *Ekologiia i Ugolovnaia Pravo: Poisk Garmonii,* Krasnodar: Kuban State University Press, 2011, 575.
7 V.I. Bulatov, *Rossiia: Ekologiia i Armiia: geopoliticheskie problemy i voenno-oboronnye deiatel'nosti,* Novosibirsk: TSERIS, 1999, 31.
8 G. Zherebkin, *Otvetsvennost' za nezakonnuiu rubku lesnykh nasazhdenii,* World Wildlife Fund, Vladivostok: Apel'sin, 2011; M.P. Goncalves, et al., *Justice for Forests: Improving Criminal Justice Efforts to Combat Illegal Logging,* World Bank Study, Washington DC, March 2012.
9 "The first stage of Amur tiger monitoring in the south of Russian Far East completed," WWF-Amur Branch, RFE, http://www.wwf.ru/resources/news/article/eng/9112 (accessed 28 March 2012).

1 Envisaging environmental crime in Russia

Past and present realities

Sally Stoecker and Ramziya Shakirova

Для экологии важны не только чистые реки, но и чистые руки.[1]
Environmentalists care not only about clean rivers, but also clean hands.
Vladimir Artiakov, Governor of Kuibyshev

Environmental crime in Russia, as in many countries, is a multi-faceted, intractable, and health and habitat-degenerating dilemma. Environmental crimes, like most crimes, are latent and therefore pose challenges for law enforcement to apprehend perpetrators in the act of poaching, felling timber, mishandling toxic substances, or smuggling enriched uranium. Legal and administrative mechanisms to cope with environmental crimes are often inadequate, ignored, or non-existent. For example, shortcomings of anti-corruption and environmental protection laws impede proper investigation and prosecution of crimes against nature; a regulatory culture of non-compliance and non-enforcement persists;[2] frequent reorganisation of environmental oversight responsibilities and structure by state agencies.[3] Interagency collaboration between state and regional nature protection agencies (*rosprirodnadzor*) and law enforcement bureaus is lacking. In addition, priority is often placed on investigating economic and violent crimes over environmental crimes, with little to no priority given to crimes against humanity. All of this inhibits the prevention or prosecution of environmental crimes that harm human health as a result of air, water, or soil pollution, radioactivity, to the the loss of habitat for numerous types of wildlife as forests are denuded by criminal loggers.

Globalization has brought with it new ways of conceptualizing criminal activity. Seldom confined to a country's own territory, many crimes spread across borders, involve actors in different countries and continents, can be planned via the internet, and can be carried out with the aid of mobile phones, hence environmental crimes can be transnational. Rob White, in his recent book on this topic, defines transnational environmental crimes (TEC) as "crimes that involve cross-border transference and global dimension, related to pollution of air, water, and land and crimes against wildlife (including illegal trade in ivory as well as animals)".[4] TEC also extend to crimes against humanity, which are often the fault of the government or corporations that either exploit, or sanction the exploitation of, the air, water, and soil.

In Russia, however, one key element that is limited in most discussions of environmental crimes and accidents is the voice of the people. Traditionally, Russia has shunned public dissent or debate over governmental programs. Civil society is still developing and activities such as investigative reporting can be a dangerous endeavor. We recall that even General Secretary Mikhail Gorbachev, who permitted debate and openness (*glasnost'*) in the mid-1980s to an unprecedented degree, was reluctant to share news about the nuclear explosion in reactors in Chernobyl, Ukraine, in April 1986. As a result, the time lag in announcing the travesty and enlisting foreign assistance cost lives and infuriated surrounding regions and countries that were affected by the radiation as it moved across the continent.

To be fair, at the same time, other countries also are responsible for the pollution of Russia (and the whole planet), because many countries ship their nuclear waste to Russia. Although this is legal and the Russian government has agreements for the importation and burial of nuclear waste, it is still of very high concern to the Russian population,[5] just as the proposed Yucca Mountain disposal facility in Nevada has resulted in public outcry and political wrangling for more than ten years in the United States.[6] Factories in China have dumped toxic chemicals into areas of the Russian Far East, contaminating Russian rivers.[7] In March, 2011, the earthquake and tsunami that caused the explosion of the nuclear power complex in Fukushima, Japan is also said to have affected neighboring countries with radiation. In our interdependent world today we are all responsible for the global environment.[8]

However, unlike most other countries, the sheer size and expanse of the Russian Federation facilitates, and indeed invites, the perpetration of environmental crimes. Russia's geographic expanse also presents challenges for law enforcers responsible for investigating crimes throughout Russia's vast and, in the winter months, often impenetrable territory. Russia is the largest country in the world, nearly one-sixth of the world's land mass, almost twice the size of the United States, and is home to the largest boreal forests (taiga) in the world, spanning 12 million square kilometers.[9] Unique, but endangered, species of flora and fauna thrive in the multitude of Russian forests and hundreds of inland waterways, the deepest freshwater lake in the world, and access to the Pacific, Arctic and other seas. Russia straddles Asia and Europe, and shares borders with numerous countries, including China to the east, Finland to the west, Kazakhstan to the south, and the Arctic Ocean to the north. The Transsiberian Railway traverses 9,259 kilometres or 5,753 miles, from the eastern city of Vladivostok to the northwest capital of Moscow–a journey of its whole length takes eight days and spans eight time zones.

Governing a country of this scale and scope is a daunting task. When the Soviet Union was captive to "totalitarian" ideology and the Communist Party ruled the people, strict border and passport controls and a array of Communist Party organizations and cells throughout the USSR kept the citizenry in obeisance, under control and in place. With the demise of the USSR in 1991, the structure of the country changed gradually as the 15 former Soviet republics became countries in their own right. The Russian Federation was constituted in 1993 and today consists of 83 federal subjects and eight federal districts.[10] Federal districts were introduced in 2001 by President Putin as means of extending federal power over distant regions.

The envoys of the districts are appointed by the president and serve as a liaison between the federal subjects and the federal government. In spite of these efforts to oversee the districts with special envoys, the expanse of Russia and the proliferation of subjects is too great for any meaningful control over their environmental protection to materialize. The federal subjects or "regions" must assume responsibility for their own environmental protection and should have enough regional authority to do that. Local governments at municipal level also should have enough means and power to be able to effectively organize and control environmental issues. Former Russian President Dmitri Medvedev stated in a recent meeting on environmental issues: "It is wrong [for the government] to take away control over environmental affairs from municipalities, because that is where the [majority] of people reside—neither in Moscow nor in large cities, but in municipalities."[11]

Over the decades, the mythology surrounding Russia's limitless and malleable resources has lead to depleted and poisoned soil, air, and water in an all-out quest to exploit natural resources: drilling for oil and gas, clearing forests for railroads, mining quarries for diamonds, diverting rivers to irrigate crops, and many others.[12] In addition, the disposal of tons of nuclear and chemical waste remains a conundrum, yet Russian authorities have entered into lucrative agreements with many countries to buy their depleted uranium for disposal in Russia. In 2001, President Vladimir Putin signed legislation allowing for the import of radioactive waste and spent nuclear fuel (SNF) from other countries—raising concerns by many Russian citizens that these wastes will contaminate the country further.[13] Russia has been the recipient of a total of some 1,298 pounds of highly enriched uranium fuel from Bulgaria, the Czech Republic, Germany, Latvia, Lithuania, Poland, Romania, Serbia, Uzbekistan and Vietnam. Russia is also transporting SNF within its own borders. In 2011, SNF from a nuclear facility near St. Petersburg was transported to Sosnovy Bor in Central Siberia by train, raising many concerns about the safety of the populations residing in the territory through which the 300-train caravan travelled.[14] Coping with so much toxic and radioactive waste left over from the cold war is one of the biggest environmental clean-up challenges that Russia faces.

Legal tools for addressing environmental crime

Russia's first nature-protection laws date back to the Tsarist era in the seventeenth century and were included in legal codes (*ulozhenii*) until the Revolution in 1917.[15] In 1920, the criminal code, as it was then called, addressed the protection of trees and for the first time named forest fires a crime.[16] Two years later, the protection of trees was elaborated in article 99 of the criminal code, providing for the punishment of individuals and forest administrations (*leskhozy*) that violated the established government economic plan by hunting and fishing at prohibited times and places. The major addition to the 1926 code, in terms of the environment, was protection of marine life, as well as rivers and lakes. The misuse of the earth (*nedr*) via extraction of natural resources was deemed criminal in article 87.

The 1960 Criminal Code, signed into law by First Secretary of the Communist Party Nikita Khrushchev, expanded the code of 1926 and contained 12 norms that could be applied to environmental protection, including the violation of veterinary laws (art. 160), illegal hunting of seals and beavers (art. 164), and the pollution of reservoirs and air (art. 223). Environmental crimes were contained in the criminal code's chapter on economic crimes thereby illustrating their subordination to the latter.[17] At this time in Soviet history, the preamble to the law on protecting the environment emphasized the economic benefits to be derived from exploiting the country's vast natural resources, and "extensive" approach as opposed to an "intensive" approach that would use natural resources wisely and treat them in a way that ensured renewal and regeneration. Nature was viewed as a boundless source of raw material for industrial and agricultural production; emphasis in the law on means of protecting natural resources or human health was limited.

Nature and her resources in the Soviet state made up the natural basis for developing the economy and served as a source of continual growth of material and cultural values that provide the best conditions for work and recreation. [18]

In the 1960s few convictions for environmental crimes were handed down. In 1963, there were five; in 1967, twelve.[19] Efforts to investigate and prosecute air and water pollution crimes were stymied by the Soviet emphasis on fulfilling and over-fulfilling industrial and agricultural five-year plans. Poorly regulated and neglected nuclear and industrial complexes emitted radioactivity and toxic levels of air pollution, and grandiose and unsustainable river diversion schemes all were byproducts of the Soviet command-administrative system engineered by General Secretary of the Soviet Union Joseph Stalin, and were geared toward fulfilling production plans set by the state irrespective of the impact on the environment or human health.

Decades later, following the collapse of the Soviet Union and the breakup of the Soviet republics into indivdual states, it was incumbent upon Russia to adopt a new set of codes to guide Russian jurisprudence. The first President of the Russian Federation, Boris Yeltsin, presided over the drafting and implementation of a new Criminal Code and Criminal Procedural Code adopted in late 1996 and introduced January 1, 1997.[20] Like the codes before it, there were positive and negative aspects of the new environmental protection legislation. Environmental crimes assumed a higher status in the 1996 code as a separate chapter of the code and were not subsumed under economic crimes.

Today, many leading legal scholars in Russia agree that the laws are incomplete and need further improvement and elaboration. A conference held in Gelendzhik (Kuban) in 2011 brought numerous legal scholars together to discuss criminology and criminal law in the realm of environmental protection. Papers presented at the conference were published in a large volume that sheds tremendous light not only on concerns about the environment and the criminal legislation for protecting it, but also on the large number of scholars who are thoughtfully, actively pursuing research on one of the most important topics of today.[21]

Many scholars referred to ongoing violations of article 42 of the 1993 Russian Constitution that guarantees a healthy environment to all citizens. Certainly, it can be difficult to ascertain or to actually "see" health-threatening contaminants in the air and in the water until it is too late.[22] Regular monitoring of water quality is imperative, but apparently not carried out in all regions. T.V. Raskina of the Procuracy Academy in Moscow points out that *not a single criminal case involving air pollution's effect on human health has been opened in the past five years*; however, in a 2010 public opinion poll, 83 percent of the respondents fear the impact of polluted water and air more than any other environmental crime.[23] Out of all crimes registered annually, only 1.0–1.5 percent are environmental crimes: 50 percent of which are for illegal logging, 35 percent for illegal poaching of marine life, 4 percent for illegal hunting.[24] Yet none of these crimes are directly harmful to human health. Further, Raskina cites several clinical studies that illustrate the impact of pollution on Russian babies and children:

> From 40 to 70 percent of children in Russia are born with developmental pathologies due to harmful environmental influences ... 350,000 persons die per year as a result of environmental pollution.[25]

Dr. Alexei Yablokov, former head of the Interagency Ecological Security Committee of the Russian Security Council and advisor to the late President Boris Yeltsin, in his article presents the following statistics:

> Every year in Russia, approximately 35,000 people die as a result of car accidents, 40,000 from alcohol poisoning and 490,000 from environmental-related diseases (data taken from WHO in 2004). Furthermore, experts claim that about half of Russia's 180,000 miscarriages per annum are due to environmental causes.[26]

Highly polluted Russian cities see an average morbidity rate of 40 percent; heavy metals in the soil lead to diseases of the endrocrine and urogenital systems, and heightened concentrations of carbon monoxide, chromic anhydride, chromic acid, lead, and styroles lead to reproductive difficulties.[27]

There is general agreement among legal scholars and criminologists that the punishments for environmental crimes are too mild; perpetrators do not fear getting caught. In many cases, the offenders' illegally gained profits from logging or poaching far outweigh the minimal fines or prison terms that may be levied against them by the law enforcement authorities. Some crimes bringing harm to the environment and to human health also have lenient punishments.[28] Leading environmental legal scholars Olga Dubovik and Tatiana Rednikova stress the poor performance of nature protection agencies and the failure to punish the majority of criminals who harm the environment.[29]

Blame can also be placed on the law-enforcement establishment. Are they working proactively to investigate crimes? Do they have the technology to pursue complex cases? Or are some of them selling out to corruption by taking bribes to "turn a blind eye" to crimes against nature?

Corruption is undoubtedly at work in almost every environmental crime involving the exploitation of natural resources, such as oil and gas extraction and timber-processing as well as activities surrounding them, such as subsidies, auctions, taxation, and licensing.[30] According to Raskina:

Professional environmental crime, coalescing with corrupt state bureaucrats who are responsible for making decisions with regard to natural resource regulation, have begun to take on the traits of organized crime.[31]

Indeed, to many observers of Russian crime, it already has.

Environmental crimes against humanity

For decades medical and demographic authorities and scientists have bemoaned Russia's unrelenting demographic crisis. According to recent estimates from the UN Development Program, Russia's population dropped by 12 million people in the last 16 years and if changes are not made to improve the situation the population could fall by 11 million more people by 2025.[32] That would reduce the population from approximately 141 million to 130 million. Russia's demographic crisis is a result of intersecting health and longevity trends, including abnormally low life-expectancy rates, especially among men; high infant mortality rates; declining female reproductive capacity, and congenital birth abnormalities.[33] Traditionally, much of the blame for Russians' poor health and high rates of morbidity and mortality[34] has been placed on non-environmental factors, including alcoholism, smoking, drug use, suicide, and work-place accidents.[35] Environmental degradation and its influence on health is often more difficult to measure. [36]

High rates of disease were prevalent in the Soviet era when the command economy drove factories to meet ambitious and often unrealistic industrial and agricultural quotas, irrespective of the environmental impact.[37] Murray Feshbach, an esteemed demographer and analyst of health in the Soviet Union, wrote volumes about chronic diseases that plagued the population over the course of several decades.[38] Together with Russian epidemiologists and demographers, Feshbach compiled alarming statistics on high mortality rates, low reproductive rates, numerous varieties of cancers, lung diseases, and tuberculosis stemming from environmental pollution, among other factors. We believe that among the most influential factors in determining health, longevity, and quality of life in Russia are those that lie in the environment: namely, water, soil, and air.

A 2004 study conducted by Russia's environmental expert, Aleksei Yablokov, revealed that there is *not a single region in the entire country* left untainted by radioactive, industrial, and/or agricultural pollution.[39] Yablokov asserts that the Russian government consistently behaves in a criminal fashion because it violates the rights of its citizens to a healthy environment as guaranteed by article 42 of the Russian Constitution.[40] According to Yablokov, new contaminants are being added to the pre-existing ones because the state agencies that protected the environment were liquidated in 2000. Paraphrasing the actions and excuses of the Russian government, Yablokov states:

Ecology is a matter for rich countries; we will take up ecology when Russia becomes rich; in order to become rich, Russia must exploit its natural resources; in order to prevent the people from worrying about pollution, we should suppress ecological information.[41]

Despite the prevalence and importance of environmental influences on public health in Russia little effort has been devoted to addressing the problems that have placed Russia 132th in Trend EPI (Environmental Performance Index) produced annually by Yale University [42] among 132 countries in 2012 . According to this report, Russia is ranked 106th for failing to improve its environmental management over time.

According to official Russian statistics, 56.5 percent of all deaths in Russia occur as a result of circulatory illnesses, 14.6 percent due to neoplasms or tumors.[43] Taken together, these two illnesses account for more than 70 percent of all deaths. Official statistics reveal that populations residing in regions co-located or in proximity of industrial enterprises have elevated morbidity levels.

A team of Swedish, British, and Russian demographers concluded that air and water pollution decreased with the collapse of heavy industry in the early 1990s.[44] However, the demise of the USSR and rise of a market economy served to strengthen the role of industry in Russia, although there was certainly a pause between the regime changes, the industrial and agricultural landscapes are thriving and expanding; the affect of their air and water emissions on public health is likely as detrimental as it was under socialism. Although industrial and economic development benefits the Russian economy, most enterprises use outdated technologies and environmental monitoring of enterprises is lacking. Today the profit motive drives industrial production in the Russian market economy, instead of the five-year plan of the Soviet command-administrative system of the socialist past.

Industrial pollution, mentioned above, is a common problem. Metallurgical, chemical and petrochemical production are major growth industries in Russian regions, particularly in Central Russia where large reserves of oil and gas exist. Oil-rich regions such as Bashkortostan, Tatarstan, and Tiumenskaia oblast are among the regions with the highest rates of morbidity in Russia, as we will illustrate below. Although many military industries have "converted" to peaceful work, their radioactive waste remains embedded in soil and stored in containers and has not been properly disposed of.[45] Mining and other extractive industries are experiencing growth with little to no attention to environmental impact and sustainability. As Russian construction firms work to prepare to host the 2014 Winter Olympics in Sochi on the Black Sea, environmental watch groups report ongoing crimes and eco-violations. [46] In August of 2011, a 30-km spill of liquid waste into the Mzymta River, the main source of drinking water for the Olympic games, poisoned not only the river, but spilled over into the Black Sea as well.[47]

The environmental pollution–health nexis

Environmental pollutants affect human health in an insidious, often slow, but tragic manner. It is much easier to blame people's habits (drinking, smoking, and drug abuse) than to examine the condition of the water they drink and air they breathe and how that affects their wellbeing. Ignorance about and dismissal of environmental poisons is not just a crime; it is a crime against humanity. In order to strengthen and expand the economy and develop the country, a healthy population is required. Sick workers who are frequently absent from work or work poorly diminish the country's labor productivity at a time in history when human capital is a country's most valuable economic resource.

To illustrate more concretely the scope and types of illnesses occurring in Russia as a result of pollution we will present statistical data on pollution and morbidity levels by federal districts. Although more extensive research may be required to estimate the pollution–morbidity relationship, a comparison of the dynamics of these two indicators shows that morbidity levels and pollution are related. More research is required to determine to what degree the two are related.

The catastrophic decline in the population is explained by abnormally high death rates and morbidity as well as a drop in birthrates. The causes of morbidity must be identified and isolated and the quality of medical care must improve. As mentioned above, there are many factors that affect health, including lifestyle (alcoholism, smoking, and drug use). In addition, there are hypotheses about the harmful influence of economic shock, poor medical treatment, and others, yet far less research has been done on the impact of pollution on human health. Yablokov and others have assembled data on nuclear explosions or contamination of water and air in the former Soviet Union: Dmitrovgrad, Chernobyl, Cheliabinsk, Seversk and more.[48] All forms of pollution affect health, as well as air, water and soil.

Analysis of morbidity and pollution in federal districts

An analysis of the information provided by the Russian Federal Agency of Statistics give a general picture of the environmental situation and the morbidity level of the population according to the eight federal districts of Russia. This is not a complete picture, however, because Russian monitoring of pollution is neither strictly enforced nor recorded.

Figure 1.1 illustrates the dynamics of air pollution from stationary sources in the Federal Districts during the years 2000-2008. As the figure shows, three districts stand out: Urals, Siberian, and Volga. These three districs are the main industrial regions (Volga District produces 20.3 percent of total industrial output in Russia, Urals – 19.2 percent, and Siberian – 12.18 percent, the data for 2010).[49] The pollution illustrated here is industrial pollution and does not include pollution from transportation. In general, there has been no decrease in emissions during these eight years. In the Siberian and Urals Districts, however, there has been an increase. We should also underscore that the share of the Urals District in total emissions in the Russian Federation in 2008 is 29 percent, Volga – 14 percent, Siberian – 30 percent. Forty-three point seven percent of the population of Russia lives in these three regions (more specifically, in Volga District – 21.3 percent, in Siberian – 13.8, and in Urals – 8. 6 percent of the total Russian population).[50]

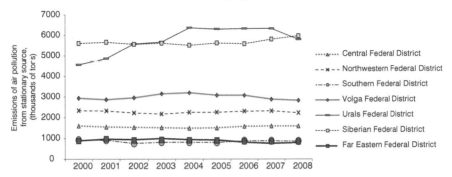

Figure 1.1 Dynamics of air pollution emissions from stationary sources, by federal district

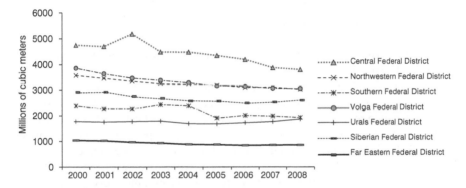

Figure 1.2 Waste discharge to waterways

Figure 1.2 shows the dynamics of discharging wastewater into waterways. The Central Federal District has the worst pollution, followed by the Volga Federal District.

Figure 1.3 shows the dynamics of morbidity level by federal district. Morbidity does not represent the total number of ill people, but rather those who were registered with this diagnosis the first time in their life. According to these data, the Volga Federal District has the highest morbidity of all. It is closely followed by Siberian, Urals, and Central Districts. The Southern Federal District is difficult to evaluate, because it has the least amount of data containing indicators of morbidity. This may be explained by the erratic organization of health care and record-keeping of morbidity in this conflict-ridden area of Russia that includes the North Caucasus. It is clear that in all districts, morbidity has been on the rise for the past eight years.

We have also analyzed morbidity by type of illnesses, and in all groups the morbidity in the Volga region is higher than average for Russia. The Volga District is where more than one-fifth of the Russian population live. The types of disease that are most harmful to health and may affect demographic trends are related to the circulatory system and the growth of tumors, as they are the main causes of

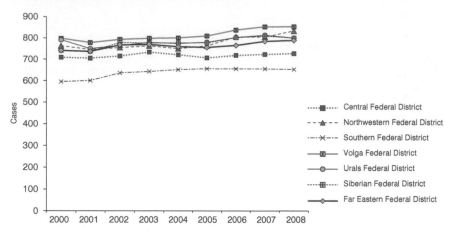

Figure 1.3 Morbidity per 1,000 persons

death in the Russia. Morbidity for these types of diseases are highest in Volga and Siberian Federal Districts.

Although morbidity rates can be influenced by quality of health care and/ or access to care, or income level, in Russia, it is clear that the industrial and highly polluted regions of the Volga, Siberian, and Urals Districts have the highest rates and therefore cannot be ignored much longer by policymakers. In some of these Federal Districts, where morbidity is higher than average there are regions (federal subjects) where the situation is even worse. Some of the regions that have the highest levels of morbidity in Russia and in almost every category of disease include Orenburgskaia Oblast, Republic of Chuvashiia, Republic of Tatartsan, Republic of Bashkortostan, Ulianovskaia Oblast, Tiumenskaia Oblast, Kurganskaia Oblast, Chelyabinskaia Oblast, Altaiskii Krai, Republic of Altai, and Omskaia Oblast. In our opinion, the environmental factor is among the main causes of poor public health and immediate policy and public health responses are essential. These are highly polluted and densely populated regions.

Conclusions and recommendations

To conclude, the protection of the environment is one of the most urgent policy problems facing Russia and almost every other country on the face of the earth. A sound environment provides good health, a productive work force, an abundance of flora and fauna, potable water and food security. Bleak demographic trends can be reversed with the protection of the environment and the regeneration of currently toxic regions and cities. As the Russian constitution provides, all citizens have the right to a clean environment and to information regarding actions that harm the environment and their own wellbeing.

To ensure this, an *environmental surveillance framework* should be established to involve civil society, proactive nature protection and environmental enforcement agencies, and governmental stakeholders representing communities, municipal, regional, and federal authorities in monitoring air and water quality and the industries that contaminate them; reporting and acting on threats or damages to the human habitat. Censorship regarding environmental accidents and crimes must not be tolerated any longer. Participation by the citizenry is essential in assisting government and private structures to identify and address environmental matters. Investigative journalism can probe deeply into this realm and reveal corruption and crimes that are contributing to environmental degradation in Russia and assist law enforcers in their investigations and prosecutions. Stronger laws are required to punish perpetrators of environmental crimes, such as poaching and illegal logging. As has been shown, the profits earned from the crimes are usually much higher than any fine violators may be required to pay for violating the law; laws lack deterrence value. Stronger laws with harsher punishments combined with proactive and dedicated law enforcement investigators will have positive impact.

Russia cannot wait until it is "rich" to address the impact of environmental pollution on its citizens' health and longevity. If Russia waits much longer, its population will be too unfit or too small to keep the economy running. As the Russian government claims demographic challenges as a top priority, it should also prioritize environmental governance as the two are interdependent.

Notes

1 "Meeting with President Medvedev about developments in the Samara Region, 15 March 2012", http://eng.kremlin.ru/news/3549 (accessed 21 April 2012).
2 M. Ivanova, "Environmental crime and punishment in Russia: Law as reason for breach," *Proceedings of Berlin Conference on the Human Dimensions of Global Environmental Change*, Potsdam Institute for Climate Impact Research, 2002, 77.
3 O. Dubovik and T. Rednikova, "Environmental crime in the Russian Federation: Status, trends, and perspectives on how to confront it," draft report for TraCCC, July 2010.
4 R. White, *Transnational Environmental Crime: Toward an eco-global criminology,"* Abingdon: Routledge, 2011, 3.
5 " Украина выполнила обещание вывезти в РФ обогащенный уран" http://eco.ria.ru/business/20120322/602655048.html (accessed 22 March 2012).
6 P. Behr, "The Yucca Mountain nuclear waste site lives on – in NRC and Capitol Hill infighting," *New York Times*, 5 May 2011.
7 S.I. Levshina, N.N. Efimov and V.N. Bazarkin, "Assessment of the Amur River ecosystem pollution with benzene and its derivatives caused by an accident at the chemical plant in Jilin City, China," *Bulletin of Environmental Contamination and Toxicology* 83/6, 2008, 776–779, DOI: 10.1007/s00128-009-9798-1
8 A. Makhijani, *Post-Tsunami Situation at the Fukushima Daiichi Nuclear Power Plant in Japan,* Institute for Environmental and Energy Research, Takoma Park, MD, March 2011.
9 U. Runesson, Faculty of Natural Resources Management, Ontario, Canada, http://www.borealforest.org (accessed 12 March 2012).
10 Constitution of the Russian Federation, Chapter 3, Article 65, Structure of the Russian Federation, http://www.constitution.ru/en/10003000-04.htm and http://www.enotes.com/topic/Federal_subjects_of_Russia (accessed 9 March 2012). An outline of Russia's federal districts is found on the European Union website: http://www.eu-russiacentre.org/russia/political-system. (accessed 15 March 2012).

11   L. Chizhova, "President and ecologists seek a common language," Radio Svoboda, 15 March 2012. http://www.amic.ru/news/175655/ (accessed 17 March 2012).
12   Norges Naturvernforbund, "Evironmental Issues in Russia,"http://naturvernforbundet. no/international/environmental-issues-in-russia/category930.html (accessed 16 April 2012).
13   Bellona Foundation, "Import of nuclear fuel to Russia," http://www.bellona.org/ subjects/Import_of_nuclear_fuel_to_Russia (accessed 8 March 2012).
14   Bellona Foundation, "Russia's plan to move spent nuclear fuel to Siberia raises safety concerns-and fails to solve the mounting waste problem," http://www.bellona.org/ articles/articles_2011/Siberia_snf (accessed 12 March 2012).
15   A.G. Kniazev, D.B. Churakov, A.I. Chuchaev, *Ekologicheskaia Prestupnost'*, Prospekt, Moscow, 2009, 5.
16   Ibid., 8
17   "Ugolovnyi kodeks RSFSR ot 27.10.1960." www.lawrussia.ru/bigtexts/law_3558/ index.htm (accessed 5 March 2012).
18   Preamble "Ob okhrane prirody v RSFSR," Federal law of the RF from 10 July 2001, no. 93 http://base.consultant.ru/cons/cgi/online.cgi?req=doc;base=LAW;n=121971;fl d=134;dst=100528 (acessed 21 April 2012).
19   Discussion with G. Zherebkin, former prosecutor and legal adviser, World Wildlife Fund, Amur Branch, Vladivostok, 2–3 March 2012.
20   Criminal Code of the Russian Federation, http://*www.ugolkod.ru/* (accessed 5 March 2012).
21   V.R. Koniakin and M.L Prokhurova, eds., *Ekologiia i ugolovnoe pravo: poisk garmonii*, Kuban State University Press, Krasnodar, Russia, 2011; materials from the Scientific-Practical Conference in Gelendzhik, Russia, 6–9 October 2011.
22   Constitution of the Russian Federation: Chapter 2. Rights and Freedoms of Man and Citizen, Article 42: "Everyone shall have the right to favourable environment, reliable information about its state and for a restitution of damage inflicted on his health and property by ecological transgressions." And article 58: "Everyone shall be obliged to preserve nature and the environment, carefully treat the natural wealth." http://www. constitution.ru/en/10003000-01.htm (accessed 5 March 2012).
23   T.V. Raskina, "Environmental crime: contemporary status and paths for countering it," in *Ekologiia i ugolovnaia prava: poisk garmonii*, 67–72, Krasnodar, Russia: Kuban State University Press, 2011.
24   Ibid.
25   T.M. Maksimova, "Issues surrounding the evaluation of environment on health and the modernization of statistics," *Na puti k ustoichivomu razvitiiu Rossii: Biulleten'39*, 2007, 12, in Raskina, "Environmental crime," 69.
26   A. Yablokov, "The environment and politics in Russia," *Russian Analytical Digest*. 79/10, 2
27   Russian State report "On the sanitary-epidemiological situation in the Russian Federation in 2004," Federal Center of Hygeine and Epidemiology, *Rospotrebnadzor*, 200 in Dubovik and Rednikova, "Environmental crime in the Russian Federation."
28   N.A. Lopashenko, "Russia's Criminal Policy and Combating Environmental Crimes: Harmony of Chaos or Chaos of Harmony," 39-47 in R. Koniakin and M.L Prokhurova, eds., *Ekologiia i ugolovnoe pravo: poisk garmonii*, Krasnodar, Russia: Kuban State University Press, 2011.
29   Dubovik and Rednikova, "Environmental crime."
30   R.O. Dolotov, "The Influence of Corruption on the Environmental Situation in the Region." 103-108 in R. Koniakin and M.L Prokhurova, eds., *Ekologiia i ugolovnoe pravo: poisk garmonii*, Krasnodar, Russia: Kuban State University Press, 2011. See also E. Jelenska, "Landfills in Russia: Precarious grounds for waste management opportunities," *Frost and Sullivan*, 22 November 2010, http://www.frost.com/prod/ servlet/market-insight-top.pag?docid=216 (accessed 5 March 2012).

31 Raskina, "Environmental crime," 70.
32 www.undp.ru/documents/NHDR_2008_Eng.pdf (accessed 27 January 2012).
33 M. Feshbach, "Behind the bluster, Russia is collapsing," *Washington Post*, Opinions, 5 October 2008.
34 Morbidity is the relative incidence of disease and mortality is the proportion of deaths to population.
35 See, for example, O.B. Resta, "Russia's demographic decline," *Saisphere*, Johns Hopkins Univerity Press, Baltimare, MD, 2010–2011, 78–81 and Peder Wallberg et al., "Economic change, crime, and mortality crisis in Russia," *British Medical Journal*, 1 August 1998, 312–18.
36 N. Eberstadt, *Russia's Peacetime Demographic Crisis: Dimensions, Causes, Implications*, National Bureau of Asian Research, Seattle, WA, 2010, 136.
37 See D. Filtzer, "Poisoning the proletariat: Urban water supply and river pollution in Russia's industrial regions during late Stalinism," *Acta Slavica Iaponica*, 26, 85–108.
38 M Feshbach, "Ecological disaster: Cleaning up the hidden legacy of the Soviet regime," Twentieth Century Fund Press, New yortk, 1995; M. Feshback and A. Friendly, *Ecocide in the USSR: Health and Nature under Siege*, Basic Books, New York, 1992.
39 A.Yablokov, *Zelenoe dvizhenie i grazhdanskoe obshchesvto: narushenie ekologicheskikh prav grazhdan v Rossii*, KMK, Moscow, 2004.
40 Ibid., 10.
41 A.V. Yablokov, *Rossiia: zdorové prirody i liudei*, Moscow: Galleia-print, 2007.
42 Environmental Performance Index, Yale University, 2012, http://epi.yale.edu/epi2012/countryprofiles (accessed 23 March 2012).
43 *Demograficheskii ezhegodnik Rossii, 2010*, Federal'naia sluzhba gosudarstvennyoi statistiki RF, http://www.gks.ru/doc_2010/demo.pdf, p. 221 (accessed 10 March 2012).
44 P. Wallberg, et al., "Economic change, crime, and mortality crisis in Russia: regional analysis," *British Medical Journal* 317, 1 August 1998.
45 D. Bradley, *Behind the Nuclear Curtain: Radioactive Waste Management in the Former Soviet Union*, Battelle Press, Richland, WA, 1997.
46 "Russian Monitors Confirm Toxic Spill to River Near 2014 Olympics," *Environment News*, www.ens-newswire.com/ens/sep2011/2011-09-01.html (accessed 13 March 2012).
47 Ibid.
48 See, for example, A. Yablokov, *Chernobyl: posledsviia katastrofy dlia cheloveka i prirody*, Universarium, Kiev, 3rd edn, 2011; A. Yablokov, *Zelenoe dvizhenie i grazhdanskoe obshchestvo: narushenie ekologicheskikh prav grazhdan v Rossii*, KMK, Moscow, 2004; V. Lystsov, "Radiological aspects of the accident in Tomsk," *Atomnaia Energiia* 74/4, 1993, 364–67.
49 Calculated using data from Federal Agency of Statistics "Regions of Russia in 2011," http://www.gks.ru/bgd/regl/b11_14p/Main.htm (accessed 23 March 2012).
50 Calculated from *Demographic Yearbook of Russia*, http://www.gks.ru/doc_2010/demo.pdf (accessed 24 March, 2012).

2 Russia's nuclear industry and the environment

Dmitry Samokhin and Alexander Nakhabov

Environmental problems associated with nuclear energy are closely related to the history of nuclear energy in Russia, specifically with regard to the colossal amount of radioactive waste left over from many years of nuclear reactor operation. This chapter is divided into two main parts. Part one begins by exploring the recent revival of Russia's nuclear industry, showing how political will and economic prudence have contributed to an increase in support for the industry. To contextualize the next section, which focuses on the mishandling of nuclear materials, attention is turned to the basics of the nuclear fuel cycle and resulting production of nuclear waste. Next, the chapter discusses methods of disposing of nuclear waste and touches briefly on problems with waste disposal before moving to liquid waste disposal practices in Russia. The final section of part one is devoted to a survey of Russian nuclear energy experts, which explores the nature of safety and security deficits in Russian nuclear energy practices.

Part two of the chapter is devoted to norms and legal issues as they relate to Russia's nuclear industry. The first section of part two sets the stage by outlining the development of international norms as well as foreign laws and regulations on nuclear energy. The next section turns to the Soviet legacy of Russia's nuclear industry and questions of adherence to contemporary norms. This sets the stage for the next section, which focuses on current Russian environmental regulations. The chapter then notes a significant legal change that may exacerbate environmental challenges related to the storage of nuclear waste in Russia. The final section focuses on an analysis by a leading Russian nuclear scientist of shortcomings in Russia's nuclear safety standards.

Some concluding thoughts are then provided in addition to recommendations for social, technical, organizational, and legal improvements to better realize the promise of Russia's nuclear industry.

Part 1

Russia's nuclear renaissance

After the Chernobyl accident and the fall of the Soviet Union, the tempo of nuclear power plant (NPP) construction in Russia was severely curtailed. Largely due to inertia, the third unit of the Smolensk NPP (1990) and the fourth unit of the Balakovo NPP (1993) were completed, before giving way to a period of quiescence. Only in the early 2000s, after Russia had recovered from the systemic crises of the late 1990s, did atomic power station construction begin anew. First and foremost, efforts were aimed at finishing construction of mothballed nuclear power plant units with VVER-1000 reactors (Russian versions of a pressurized water reactor): the Kalinin NPP (a third unit went operational in 2004) and the Rostov NPP (1 unit operational in 2001, a second unit operational in 2010, with two more currently under construction). Additionally, construction has begun on new projects: the Baltic NPP (Kaliningradskaia oblast, two VVER-1200 units), the Beloiarsk NPP (a fourth unit with a BN-800 reactor),the Leningrad NPP-2 (two VVER-1200 units), and the Novovoronezh NPP (two VVER-1200 units).

There are many reasons for this "nuclear renaissance." Russia's need for new power generation sources to compensate for growing energy use and conservation of fossil fuels, which contribute significantly to the country's total exports, has helped to drive increased interest in the nuclear sector. Shifts in political and social contexts have also contributed to greater support for atomic energy. Russian politicians have suggested that the quick tempo of growth in the atomic energy sector will be preserved for the immediate future.

Nuclear fuel and waste production

According to Russian law, nuclear materials are defined as that which contain or are able to contribute to fissile materials. Materials not classified as nuclear but that emit ionized radiation are considered radioactive. Finally, nuclear and radioactive materials that will no longer be used are classified as radioactive waste.[1] In a majority of instances, these three groups of substances play a role in environmental crimes related to the mishandling of nuclear energy resources.

The atomic energy industry and military–industrial complex are most actively involved in handling nuclear materials and radioactive substances. Depending on who is involved, approaches to resolving mishandling of nuclear materials are substantively different.

Mastery of nuclear energy by humans initially took place during the process of developing nuclear weapons. Figure 2.1 depicts the production cycle for creating nuclear weapons. Nuclear materials or radioactive substances are handled at every stage of the production cycle. Civil atomic energy (here meaning nuclear transportation for civilian and military purposes) has inherited most of the features of the weapons production cycle. Figure 2.2 shows the open nuclear fuel cycle as well as possible variations for a closed nuclear fuel cycle.

The quantity of nuclear materials used at each stage of the production cycle and the corresponding volume of radioactive waste differs substantively for military and civilian purposes. This is because enrichment levels for weapon-grade nuclear materials are far greater than enrichment levels for civilian energy needs. For example: 169 tons of mined uranium (0.72 percent ^{235}U) yields 20 tons of 4 percent enriched uranium for use in a light-water reactor and leaves 149 tons

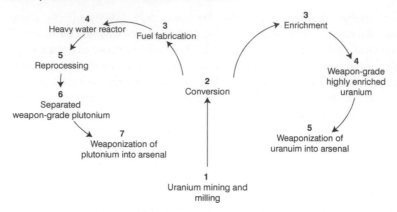

Figure 2.1 Nuclear weapons cycle (source: "The nuclear weapons cycle," The Institute for Science and International Security, http://www.isisnucleariran.org/assets/images/weapons-big.gif).

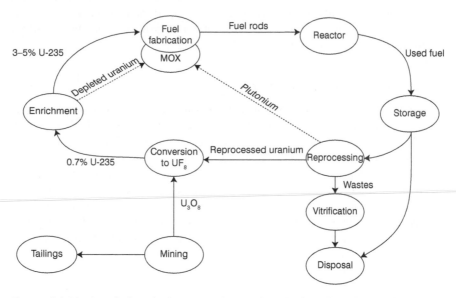

Figure 2.2 Nuclear fuel cycle (source: "The Nuclear Fuel Cycle," The World Nuclear Association, http://www.world-nuclear.org/info/inf03.html).

of radioactive waste (0.3 percent [235]U); 111.5 tons of mined uranium are needed to get 500 kg of highly enriched weapons-grade uranium (90 percent [235]U), leaving 111 tons of radioactive waste.[2] As a result, the share of radioactive waste from the uranium enrichment process amounts to 88 percent and 99.56 percent for civilian and weapons-grade nuclear material, respectively.

Depending on its physical state, radioactive waste can come in liquid, solid, or gas forms and can be separated by activity levels: low, medium, and high. The most common form of highly radioactive waste is processed nuclear fuel, which is

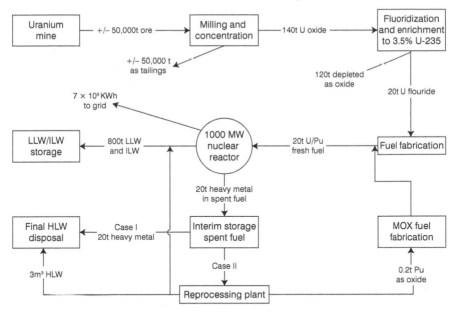

Figure 2.3 Nuclear fuel cycle for a 1,000 mw light-water reactor (source: Peter W. Beck, "Nuclear energy in the 21st century: examination of a contentious subject," *Annual Review of Energy and the Environment*, 24, 1999, 114)

created during the processing of glassified radioactive waste after processing and other radioactive materials found within the nuclear reactor. The quantity of this waste depends on reactor type. Waste-handling requirements, including disposal, vary by country. For example, a 1,300 megawatt (MW) pressurized water reactor in Germany produced roughly 60 m³ of low-level and mid-level radioactive waste and 26 tons of spent nuclear fuel. After this reactor was decommissioned, 5,700 m³ of low-level radioactive waste was left. Over this reactor's 35 years of service, nearly 300,000 m³ of waste was produced and disposed of.[3] Figure 2.3 is a diagram of the uranium fuel cycle for a nuclear power plant with light-water reactors, each which produce 1000 MW of electric power.

Mishandling of nuclear material

Most uranium is extracted by open pit or subterranean mines. With the exception of a few deposits in Canada, uranium content in mined ore does not exceed 0.5 percent. As a result, large quantities of ore must be mined to obtain enough uranium for nuclear energy purposes. During the process of uranium mining, residual waste consisting of sand and radioactive nuclides is produced, usually in a higher concentration than normal waste. This material presents a threat to the surrounding environment because after a mine closes the material continues to serve as a source of radon and radioactive water. Leakage from a waste disposal site in eastern Germany (built by the company Wismut, formerly of the German

Democratic Republic) reached levels upward of 600,000 m³ annually, only half of which was returned to the disposal site. Wismut's uranium mines in Germany's Schlema Aue area have a total volume of 47 million m³ and cover 343 hectares.⁴

Too often waste dumps can be found in close proximity to residential neighborhoods. Moreover, materials from waste dumps are mixed with gravel or cement to build roads, meaning that concentrated radioactive materials are sometimes distributed across a wide geographic area.⁵ By 1982, the USA alone had accumulated 175 million tons of waste, which included depleted uranium and radon. Although special facilities have been built to contain the waste, earthquakes, heavy rain or flooding can seriously damage these facilities. For example, in 1977 50,000 tons of liquid waste and several million litres of contaminated water leaked near Grant, New Mexico. Additionally, nearly 1,000 tons of liquid waste and 400 million tons of contaminated water leaked near Church Rock, New Mexico two years later.⁶

A large volume of contaminated water used in the uranium mining process is released into rivers and lakes. Wastewater from mineral deposits at Rabbit Lake (Canada), for example, caused massive amounts of uranium to sediment in Hidden Bay and the Wollaston River. In 2000, uranium content of the sediment was eight times higher than natural levels and by 2003 uranium levels had more than doubled those of 2000. Wismut's area of operations in Germany contained radium and uranium at levels 100 times greater than normal.

A mine's ventilation shaft, which reduces health risks for miners, releases radioactive dust and radon into the atmosphere. At Wismut's Schlema-Alberodamine, for example, an estimated 7,426 m³ of pollution with an average radon volume of 96,000 Bq/m³ (Becquerel – a unit of radioactivity) was released.⁷

Other uranium extraction technologies, such as solution mining, do not require the removal of rocks from the mining site. At Konigstein (former GDR), nearly 100,000 tons of sulfuric acid were used for extraction. After the Konigstein mine was closed, 1.9 million m³ of rock and acid were left in the ground, while 0.85 million m³ of this toxic mixture can be found in areas where leaching and processing companies operated. This mixture contains a dangerously high concentration of contaminants (cadmium levels are 400 times greater than the norm, arsenic at 280 times, nickel at 130 times, and uranium at 83 times greater than acceptable norms).⁸ In particular, this mixture poses serious danger to groundwater sources. For example, 3.7 million tons of sulfuric acid were pumped into a mining site at Strass near Ralsko in the Czech Republic, leaving 28.7 million m³ of contaminated liquid at the leaching site, covering 5.74 km². Moreover, this contaminated liquid spread through the leaching site both horizontally and vertically, posing contamination threats to 235 million m³ of groundwater over 28 km².

After extraction, uranium ore is sent to uranium-processing plants and from there to enrichment plants. At this stage, waste usually takes a liquid form, which is pumped into artificial reservoirs for final disposal. Liquid waste contains an insignificant amount of uranium as well as all elements of the ore, including heavy metals, arsenic, and chemicals added during the extraction process. The quantity of waste produced here effectively equals the quantity of mined ore, because the extracted uranium represents a insignificant share of the ore's overall mass. The largest artificial reservoir in the world is located in Ressing (Namibia) and contains more than 350 million tons of waste material. Similar reservoirs in the United States and Canada hold up to 30 million tons, while reservoirs in eastern Germany hold 86 million tons.[9] In a number of instances, nuclear waste was simply dumped in the open without any control measures. In Gabon, where the French company Cogema has mined uranium since 1961, nuclear waste was dumped freely until 1975. Estimates suggest that nearly two million tons of waste were released into the environment, contaminating ground water and river sediment. When mining ceased in 1999, radioactive waste covered an erosion-prone thin layer of soil in the mine's operational zone.

After enrichment, uranium is sent to a fuel-fabrication plant, which is the last stage before it is ready to be used for atomic energy. At this point, processed nuclear fuel is held in a pre-reactor containment facility before being transported to the spent nuclear fuel processing plant. At the spent nuclear fuel processing plant, radioactive emissions are inevitable during fuel dissolution, plutonium and uranium separation, and storage processes. Despite safety and prevention systems, radionuclides are released into the atmosphere. Radiation produced by reprocessing plants at Sellafield (UK) and La Hague (France) exceed that produced annually by light-water reactors.[10] Radiation from reprocessing plants spreads to neighboring countries and liquid radioactive waste is eventually dispersed into oceans as well.

Ultimately, nuclear weapons production goals favor obtaining the greatest quantity of highly enriched uranium or plutonium as quickly as possible, leaving environmental protection as a very low priority. This situation is further complicated because most production cycle stages take place, for technical convenience, at the same location.

In the United States, weapons-grade plutonium was produced at the Hanford site (Washington state), which included nuclear fuel production plants, eight industrial plutonium reactors, one dual-use reactor, four isotope divider plants, and one plant for final plutonium processing. The Hanford site currently has significant amounts of radioactive waste from a partially destroyed disposal facility, some of which has contaminated the nearby Columbia River and has seeped into soil and groundwater.[11]

Nuclear waste disposal

At the dawn of atomic energy age, radioactive waste was largely disposed of in the ocean, although in a few countries waste was housed in long-term above-ground disposal facilities. The USA in 1946 and the UK in 1949 used the ocean disposal method, which was subsequently employed by a number of other countries. By all accounts, the UK is responsible for three-quarters of all radioactive materials dumped into the ocean.[12] From 1946 to 1970, the United States dumped roughly 87,000 containers of radioactive material, with average radioactivity levels of 100,000 Ci (curies – a unit used to measure the rate of radioactive decay), as well as a reactor from *Seawolf*, a nuclear-powered submarine, into the ocean.[13] The Soviet Union began using this method in 1959 after testing its first nuclear-powered submarine *K-3*. From 1959 to 1992, the Soviet Union dumped liquid radioactive materials with radioactive levels of 20,600 Ci and solid radioactive materials with levels of 2.3 million Ci into the Arctic Ocean. Accordingly, radioactivity levels in the seas near Russia's Far East are between 12,300 and 6,200 Ci. An accident on Soviet nuclear submarine *K-27* in 1968 led to a decision to scuttle the submarine and its flooded nuclear reactor compartments in the Arctic Ocean, while a loss of coolant accident on the Soviet icebreaker *Lenin* resulted in nuclear materials being dumped into the Arctic. Disposal of radioactive materials by the Soviet Union and now Russia did not and does not follow International Atomic Energy Agency (IAEA) recommendations. The share of radioactivity released in the Arctic Ocean over the thirty years of Russia's atomic fleet is only 3.3 percent of total radioactivity released by all countries.[14]

The IAEA initially sanctioned oceanic disposal of radioactive materials, albeit with a number of conditions to which many countries failed to adhere. Since 1993, however, the London Convention on Nuclear Dumping has prohibited disposal of radioactive materials in oceans. Repositories are currently being built in several countries for final disposal of low- and mid-level radioactive waste. Onshore temporary containment facilities for liquid nuclear waste are far from satisfactory. Four onshore nuclear waste repositories on the Kola Peninsula (located in northwest Russia) and in the Russian Far East have reached capacity, despite unsatisfactory technical conditions. Specifically, walls and fuel assembly claddings have been damaged at these facilities, which has led to leakage of radioactive materials into the surrounding environment.[15]

In Russia, no final disposal facilities currently exist for high-level radioactive waste and spent nuclear fuel. US authorities completed preparations for a high-level radioactive waste repository in a salt mine near Carlsbad, New Mexico, in 1999 and the repository continues to be in use. The purpose of final disposal in deep geological formations is long-term isolation of radioactive waste from humans and the environment. All over the world, this principle is applied for high-level radioactive waste and long-life, mid-level radioactive waste. Today, many consider waste disposal in special mine shafts to be the best method. The Yucca Mountain site in Nevada and Finland's AES Olkiluoto NPP site are the only facilities to have made significant progress in this area.

A fundamental feature of advanced nuclear energy technologies and innovative closed fuel-cycle technology, as compared to traditional thermal power, is the retention of the vast majority of harmful radioactive materials at all stages of nuclear energy production, until the final disposal of radioactive waste. In the past decade, however, the problem of handling radioactive waste has been given scant attention in many countries. This includes radioactive waste from defense-related activities as well as waste from atomic power stations and other nuclear-powered enterprises. In the meantime, radioactive waste continues to accumulate. By 2003, 415 million m³ of liquid radioactive waste (99.5 percent low-level) and 79 million m³ of solid radioactive waste (99 percent low-level), and 14,600 tons of spent radioactive fuel had been produced in Russia alone.[16] Most of this waste is produced at the Mayak Industial Enterprise facility.

In addition to these examples, the environmental impact of other nuclear operations, such as nuclear missiles and atomic submarines, must be taken into account. To date, two American atomic submarines (USS *Thrasher* and USS *Scorpion*) as well as five Soviet/Russian atomic submarines (*K-8, K-219, K-278 Komsomolets, K-159,* and *K-141 Kursk*) have been lost at sea. Nuclear reactors from Russian submarines were stopped, while the reactor compartment from the *Kursk* was raised to the surface. In most cases, however, the threat of radioactive contamination remains, due to the existence of nuclear weapons on board the sunken submarines.

Accidents at atomic power stations and nuclear fuel factories that released nuclear materials into the environment have caused significant damage. The worst accidents occurred at Windscale (UK, 1957), Mayak Industrial Enterprise (USSR, 1957), Three Mile Island (USA, 1979), Chernobyl (USSR, 1986), Fukushima (Japan, 2011).

Liquid waste disposal in Russia

In Russia, 480 km^2 of territory occupied by 22 subsidiaries of the Russian state-owned company Rosatom are contaminated by radionuclides.[17] This territory includes 376 km^2 of land and 104 km^2 of surface water. In particular, liquid radioactive waste containment facilities at the Mayak Industrial Enterprise, the Siberian Chemical Combine (SCC), and the Mining and Chemical Combine (MCC) encompass around 91 km^2 of land, of which Mayak covers 88 km^2.[18] Mayak, located in Ozersk, is one of Russia's biggest producers of nuclear weapons material. Its operations include ten industrial reactors, a radio-chemical facility for ^{239}Pu separation (Plutonium-239 is an isotope of plutonium and the primary fissile isotope used for the production of nuclear weapons), a chemical-metallurgical facility for refining plutonium, and a spent fuel regeneration facility.

Current technical guidelines allow mid-level radioactive liquid waste to be collected and disposed locally in open networks of industrial reservoirs. This approach contradicts current Russian environmental protection legislation. According to Russia's official Radiation Safety Standards-99 and Basic Sanitation Rules for Guaranteeing Radiation Safety-99, the only acceptable way to solve the problem of industrial reservoirs holding mid-level radioactive waste is to get rid of them completely. The total content of radionuclides in these reservoirs is greater than 100 million Ci, which exceeds allowable limits.[19]

SCC (located in Seversk, near Tomsk) includes an isotope dividing plant, a reactor plant with five industrial reactors, and a radio-chemical plant, where a significant quantity of low- and mid-level liquid radioactive wastes was injected into deep underground water pools. Liquid radioactive waste containing alpha-emitting radionuclides and radioactive pulp sits in industrial disposal reservoirs. The total radioactivity of accumulated radionuclides is greater than 100 million Ci.[20]

Similarly, at MCC (located at Zheleznogorsk, near Krasnoyarsk), low- and mid-level liquid radioactive waste was injected over a long period of time into deep underground water pools, as only four liquid radioactive waste disposal reservoirs were built. However, the radionuclide content at the MCC site is significantly less than that of Mayak and SCC.

Reviewing the waste disposal practices of these complexes, it is clear that a number of Russian laws (including Federal Law "On the use of atomic energy," the Federal Water Code, the Federal Law "On security for hydro-technical weapons, and Federal Law "On environmental protection") and international recommendations are being violated.

Experts survey

The release of radionuclides from containment facilities, which may lead to radioactive contamination, poses a significant danger to the environment. In order to evaluate this potential threat with respect to Russia's liquid waste disposal facilities, a group of specialists from Russia's state atomic watchdog, Gosatomnadzor, was organized to conduct a survey of Russian experts. The experts surveyed represented different bureaus from within the Ministry of Nuclear Energy and Industry, the Ministry of Science, as well as Gosatomnadzor itself. The group of specialists which conducted the survey included Y. G. Vishnevskii, V. M Iryshkin, A. I. Kislov, B. G. Gordon, I. V. Kaliberd, A. G. Leyovin, N. S. Pronkin, A. A. Smetnik, and R. B. Sharafutdinov. The survey was conducted anonymously and each respondent was assigned a temporary identification number.

Results from the experts' survey were then used to build a danger index (DI) for evaluating each liquid waste disposal facility. The experts surveyed provided the following risk assessment criteria:

1   value of total radioactivity in the light water reactor (LWR) water storage unit;
2   specific activity of the aqueous phase of the LWR reservoir;
3   specific alpha activity of the aqueous phase and sediment;
4   environmental pollution due to meteorological and hydrological conditions;
5   immediate and future consequences of groundwater radiation contamination;
6   effect of hydraulic structures on public safety and the environment;
7   probability of a fission chain reaction to radiation effects.

Review and analysis of results from the experts' survey revealed the following:

1   The B-9 waste facility (Mayak) received the highest mark on the danger index, 16.8. The second highest mark, 12.3, was given to the B-1 waste facility (SCC). The B-2 waste facility (SCC) is full, but the low quantity of radionuclides in the site's sediment meant the facility received 9.6.
2   The B-17 waste facility (Mayak) received a 7.8, due to accumulated highly radioactive materials and prior liquid radioactive waste accumulations, including alpha-active radionuclides. Furthermore, it is possible that the integrity of the B-17 structure has been compromised. The B-25 (SCC) waste facility contains long life radionuclides and therefore was given 4.83.

3   Mayak waste facilities B-3 (DI=3.5), B-4 (DI=3.0), and especially B-10 (DI=5.2) and B-11 (DI=5.8) need particular attention from a security standpoint, according to the experts surveyed.
4   Waste facilities BX-3, BX-4, PX-1, and PX-2 (all SCC) as well as 365 and 366 (MCC) all received similar marks: 3.74, 3.15, 3.6, 3.5, 2.5, and 2.6, respectively.
5   Waste water facilities B-6 (Mayak) and BX-1 (SCC) received low marks: 1.4 and 1.1, respectively.

"Fuzzy sets theory" was used during the process of organizing the expert survey as a way of clarifying shortcomings in Russian laws, which contribute to deliberate mishandling of radioactive materials. Fuzzy sets are sets whose elements have degrees of membership and can be used when information is incomplete or imprecise. The first step for assessment is developing a membership function,[21] which helps to define relative preferences with respect to legal violations that facilitate environmental crimes and the mishandling of nuclear materials. Different violations may be associated with different legal loopholes, which may affect preference ordering. Additionally, violation preferences may be attributed to information from open sources or informal contacts.

According to the expert group, nuclear waste operators are not complying with current Russian legislation and international norms regarding efforts to increase the safety and security of liquid nuclear waste facilities. Operators do not perform careful analysis of waste dumps and there are no detailed programs for decommissioning dumps at these waste facilities. For one, the annual volume of liquid waste dumped at Mayak's B-3 and B-4 waste facilities contains $(3–5)\ 10^6 m^3$ with unacceptable total radioactivity levels of 2–4 Ci/yr.

Little attention is paid to improving reprocessing technologies for low- and mid-level liquid radioactive waste. Furthermore, plans for improving the monitoring of liquid radioactive waste facilities leaks and radionuclide release into the environment have yet to be realized. Efforts to address problems at waste facilities B-9 (Mayak), B-2 (SCC), and 354 (MCC) have been largely confined to building additional liquid waste facilities for radioactive material containing high levels of long-life radionuclides. In essence, long-term safety problems related to liquid radioactive waste at these sites remain unaddressed.

Part 2

International and foreign laws and regulations relating to nuclear materials

Contemporary conceptions of global security cover many aspects, including national, social, economic, environmental, nuclear, and so forth. However, regulatory rules do not include a complete enumeration of the different types of security. This is because of the development of new technologies, which present new threats and therefore require new sets of safety precautions.

The history of atomic energy, for both military goals and peaceful purposes, demonstrates that governments have pursued atomic energy programs for scientific, military-strategic, technical, economic, and political ends. Environmental issues are more or less ignored and shelved for future generations. Additionally, scant attention has been paid to a host of legal issues related to the use of atomic energy and safe handling of radioactive material and waste. However, legal issues related to protecting the environment from radioactive contamination and the safe handling of radioactive material and waste have increasingly entered the spotlight after a series of nuclear accidents at atomic energy stations and nuclear fuel production facilities in a number of countries.

Safe handling of radioactive material remains a major problem for the nuclear industry. Special attention has been devoted to this topic only in the past 25 years, following the Chernobyl disaster. Until Chernobyl, discussions of safe handling of radioactive materials focused largely on cross-border transfers and ensuring that oceanic disposal of radioactive substances did not take place in the territory of a sovereign state. Other aspects of this issue are regulated by national laws, but not in all states currently using atomic energy.

Safe handling of radioactive materials and waste is not merely a problem for individual states, but affects the entire international community. As a result, it is particularly important to look at this problem in an international context in order to develop lessons learned and improve best practices. Furthermore, as governments have varying interests relating to the safe handling of radioactive materials and waste, many seek to completely remove radioactive materials and waste from their sovereign territory. Accordingly, international legal questions on this issue are particularly important.

The first stage in the development of atomic energy legislation (1930–1945) was marked by the expansive use of atomic energy. In this period, atomic energy was predominantly used for medical and industrial purposes. Norms were developed to protect individuals working with ionizing radiation. In addition, regulations governing the safe use of radioactive materials appeared.

The second stage of development of legislation corresponds with the beginning of the epoch of practical use of atomic energy. During this stage, from 1945 to 1953, industrial atomic energy use was mainly directed toward military ends, which meant that the world's leading nuclear powers were not inclined toward nuclear cooperation. These factors influenced the character of atomic energy legislation during this period. Examples of this influence can be seen in US and Canadian atomic energy laws from 1946, both of which established monopolies on the production and handling of nuclear materials. In 1953 and 1954, the government monopoly on atomic energy eased and international cooperation between nuclear and non-nuclear powers began to expand. A number of laws aimed at developing peaceful atomic energy were passed in the United States in 1954, in West Germany and Japan in 1955–1957, Italy in 1957, Argentina in 1956, and other countries. Legislation from this period developed in a few different ways, primarily dealing with creating regulatory agencies, defining functions, establishing licensing procedures for handling nuclear waste, developing norms for protection against ionizing radiation and developing standards for civil liability in nuclear accidents.

International atomic energy cooperation with respect to environmental protection has developed on the basis of legal-regulatory guidelines found in international conventions (such as the Vienna Convention on Civil Liability for Nuclear Damage, the 1980 Convention on the Physical Protection of Nuclear Materials, the 1986 Convention on Early Notification of a Nuclear Accident, the 1986 Convention on Assistance in the Case of a Nuclear Accident, the 1994 Convention on Nuclear Safety, and others), multilateral and bilateral agreements, as well as intergovernmental agreements on the creation of specialized international organizations (such as the IAEA), the charters of such organizations, and agreements between governments and international organizations (such as agreements concluded with the IAEA on providing state technical assistance).[22] Declarations, projects, and recommendations of international organizations, as well as methodological studies and other documents that improve understanding and encourage implementation of international norms and regulations also play an important role in helping to protect the environment.

Guidance for national atomic energy legislation can be found in the 1980 Convention on the Physical Protection of Nuclear Materials (CPPNM), which went into effect in 1987. By 1997, more than 60 countries were signatories. According to Article 2, the CPPNM governs the handling of nuclear materials used for peaceful purposes across international borders as well as the handling, storage, and transfer of nuclear material for peaceful purposes within a state. The main goal of the CPPNM was to establish broad guidelines for all states with respect to transporting nuclear materials.

The CPPNM remains one of the main documents regulating the handling of nuclear and radioactive materials and the convention's jurisdiction covers all signatories. National legislation governing atomic energy use is closely modeled on this and other international agreements. Recommendations from international organizations, such as the IAEA, and non-governmental organizations (NGOs) have played a significant role in the development of national atomic energy legislation, including criminal liability.

As the use of atomic energy has grown, so has the amount of radioactive waste. Government practices of dumping radioactive waste into rivers and seas have resulted in the serious pollution of a number of international waterways. Resolving this issue has become one of the main conditions and demands for the development of nuclear energy in developing countries. A number of governments have already passed legislation that governs conditions for providing licenses for building and operating atomic energy stations. For example, Switzerland's atomic energy law, which went into effect in 1979, provides that construction licenses will be given only if the long-term safety of radioactive waste is guaranteed by a project.[23]

In the United States, all federal agencies must follow guidelines enumerated in the National Environmental Policy Act (NEPA) for determining the human impact of nuclear energy projects. NEPA defines conditions for obtaining industrial licenses for uranium enrichment and nuclear fuel production as well as temporary radioactive waste storage at industrial nuclear energy facilities.[24]

In Canada, there are regulations that guide nuclear energy research, production, and use as well as regulations for production and use of nuclear materials that seek to eliminate unjustifiable environmental risks. Measures have been taken to ensure that standards and norms are implemented in practice.[25]

The illegal trade in nuclear materials is closely connected to environmental crimes. Interpol has played a particularly significant role in combating these crimes, having started a special nuclear division in the mid-1990s. The illegal trade in nuclear materials has been especially problematic in Europe, as many countries in the socialist bloc were reliant on atomic energy.[26] Despite the fact that in most cases the volume of radioactive materials failed to present real threats with respect to building nuclear weapons, such crimes indeed posed serious health and environmental dangers. This was exacerbated by the fact that individuals participating in the illegal trade of nuclear materials were unable to physically guarantee the necessary level of protection and safety of the nuclear materials, thus increasing risk of environmental damage.

According to Article 4 of the European Parliament and Directive 2004/35/CE of the European Parliament and of the Council "On environmental liability with regard to the prevention and remedying of environmental damage," incidents occurring at nuclear facilities are regulated by international agreements, such as the IAEA Convention on Assistance in the Case of a Nuclear Accident or Radiological Emergency and others.[27]

Operations of nuclear installations are regulated by a few different conventions regarding civil liability and are based on the principle of potential liability. In general, these conventions regulate traditional damages, but also hold governments responsible for environmental damages. However, conventions covering environmental damages are far less coordinated than those regarding traditional damages. A protocol aimed at improving the regime outlined in the 1960 Paris Convention on Nuclear Third Party Liability was passed by the Atomic Energy Agency of the OECD, in which 13 of the 25 EU states are represented. This directive does not address damages borne by nuclear weapons, which are outlined in a number of international agreements.[28]

The Soviet legacy and contemporary adherence to legal norms

It should be noted that liquid radioactive waste facilities in Russia largely operate in accordance with industry regulations, which provide guidelines for the following: proper balances for water level fluctuations; discharge limits; measures for preventing wind entrainment of radionuclides; water conservation methods; organization and conduct of environmental assessments, and so forth.

However, official industry guidelines are based on laws and regulations from the Soviet period. One of the foundational texts for current guidelines is the decree of the Council of Ministers of the USSR "On measures for improving use of and strengthening protection of the water resources of the USSR," from April 22, 1960. This decree is limited to questions involving the use of water resources. Effective water legislation, however, should not only regulate the use of water resources but also that of water facilities. For example, international water regulation norms prohibit the permanent or temporary operation of facilities producing and introducing untreated waste water into the environment. Although designed to protect water resources from contamination, waste water producing facilities are charged with the responsibility of having necessary water treatment equipment. As such, effective legislation cannot ignore issues related to waste water cleanliness levels and should include measures establishing definitions for designing and operating waste water treatment equipment. Similar problems can be found in legislation regulating the use of other natural resources.

In their current form, nuclear energy industry guidelines do not reflect contemporary safety conditions and environmental requirements, Russian legislation, international norms, or recommendations from international organizations.

Specifically, liquid radioactive waste facilities' operations violate a number of laws and regulations, including:

- failure to guarantee safe isolation of radioactive material from the surrounding environment (Federal law "On the use of atomic energy," Article 48);
- failure to guarantee protection of current and future generations as well as natural resources from contamination by radioactive waste as indicated in official atomic energy standards and norms (Federal law "On the use of atomic energy," Article 48);
- violations of prohibitions on dumping radioactive materials in waterways (Water Code of the Russian Federation, Article 104);
- failure to guarantee proper environmental protections during collection, containment, and disposal of radioactive materials (Federal law "On protecting the environment," Article 51, part 1);
- violations of prohibitions on radioactive materials disposal in surface and underground water ways, in catchment areas, in subsoil, and soil (Federal law "On protecting the environment," Article 51, part 2);
- failure to ensure that radioactive waste facilities are maintained at a minimal practicable level (Joint Convention on the Safety of Spent Fuel Management and on the Safety of Radioactive Waste Management, Article 11);
- imposition of an undue burden on future generations (Joint Convention on the Safety of Spent Fuel Management and on the Safety of Radioactive Waste Management, Article 11);
- failure to ensure feasible maintenance levels for radioactive materials (IAEA Principles for Managing Radioactive Waste, Series No. 111-F)
- failure to properly take into account interdependence of all stages radioactive waste build-up and disposal (IAEA Principles for Managing Radioactive Waste, Series No. 111-F).

Russian environmental regulations

The State Atomic Energy Company, Rosatom, is tasked with minimizing the environmental impact of Russian nuclear industry. Rosatom's environmental policy outlines the goals, basic principles, and operational activities with respect to guaranteeing environmental security as well as the current and long-term development of atomic energy for peaceful purposes and nuclear weapons.[29] Rosatom's environmental policy aims to achieve internationally mandated levels for nuclear materials, radiation, and other components of environmental security as well as the development of new environmental protection technologies.

The constitution, Russian domestic laws, international norms and agreements, and the "Environmental Doctrine of the Russian Federation" form the legal basis for Rosatom's environmental policy.[30]

The main goal of Rosatom's environmental policy is to create conditions in which Russia's nuclear industry more effectively complies with its environmental policy. Those goals include protecting eco-systems, supporting their integrity and life-support function for the sustainable development of society, raising living standards, improving public health and the demographic situation as well as guaranteeing national environmental safety.

These conditions must guarantee the following:

- environmental safety of currently active, planned, and decommissioned nuclear sites;
- early resolution of accumulated environmental problems;
- development and implementation of new, efficient, and environmentally sound atomic energy technologies and practices.

Strategic directions for achieving successful implementation of environmental policies include:

- implementing safety measures and improvement of environmental safety levels at all active and decommissioned sites under Rosatom's purview;
- addressing the problem of long-term safety in terms of handling radioactive material and spent nuclear fuel;
- improving industrial environmental protection and management oversight systems;
- improving laws and regulations on environmental safety and protection, while keeping in mind specific tasks outlined by the Federal Atomic Energy Agency;
- developing modern methods and means for complex analysis, forecasting, and management of environmental safety and nuclear radiation risks;
- improving environmental monitoring tools, analytical systems, and safety and control management;
- guaranteeing minimal readiness levels for preventing and responding to the consequences of accidents and emergency situations;
- developing and implementing new atomic energy use technologies, specifically with respect to reactors and fuel cycles, which improve environmental protection measures;

- deducing the negative impact of nuclear facilities on the human population and environment, based on complex analytical research;
- improving international cooperation in the areas of environmental protection and sustainable nuclear industry development.

Recent legal developments

Economic and political problems can be traced to a decision in 2000 to change article 50 of the Russian law "On protection of the surrounding environment," which allowed for the transfer of spent nuclear fuel from foreign atomic power stations to Russian territory for processing and storage. The Russian Ministry of Atomic Energy (now the Federal Atomic Energy Agency or FAEA) lobbied for the changes and accordingly is prepared to begin accepting foreign spent nuclear fuel. However, a significant portion of the population and representatives of environmental organizations have mounted opposition to the plans. At the moment, around 14,000 tons of spent nuclear fuel have accumulated in Russia, while the new plans include acceptance of 16,000 tons for processing and 4,500 tons for storage. For this, Russia will receive 20 billion dollars. FAEA's financial interest is clear. As FAEA representatives have noted, Russia needs funds for disposing its nuclear waste and spent nuclear fuel. Russian nuclear waste can be processed at low cost, and foreign money could finance the processing of both Russian nuclear waste and waste brought to Russia from abroad.

Storage of unprocessed Russian nuclear waste throughout the country threatens to be an environmental catastrophe. Russia needs new atomic energy stations and spent nuclear fuel from new stations must be disposed of – assisted by foreign funds. It must be said that nuclear scientists recently have disputed this approach and have argued against those who suggest that nuclear scientists supported this approach previously. Opponents of the introduction of foreign spent nuclear fuel maintain that Russia's nuclear waste is not great and that funds may be found elsewhere so as not to threaten the country's population. Funds received for processing and storing spent nuclear fuel from abroad may be misappropriated or may be directed towards addressing pressing economic problems. Questions of who will control the distribution of these funds as well as the source of funds for addressing possible accidents at new reprocessing plants for foreign spent nuclear fuel have yet to be answered.

Nuclear safety standards and technical documentation

According to Dr. Y. V. Volkov, a professor at the Obninsk Institute for Nuclear Power Engineering and specialist who has devoted his life to studying questions of the safe use of nuclear technologies, there are many ambiguities in relevant laws, definitions, and regulations that have given rise to double standards. These issues include the following:

1   The Federal law "On technical regulations" contains a definition of risk as the probability of unforeseen events occurring and the resulting damages from such events. However, guidelines for implementing this understanding are absent. As such, one risk component is outlined clearly: how to determine the probability of unforeseen events. The other aspect – damages – lacks a quantitative definition and is not quantitatively linked to the definition of probability.

There is a widely accepted definition of risk as the product of probability and damages, formally defined as $R=P\times D$. However, many specialists find this definition problematic because it fails to take into account how the probability of an event and damages from an event are functionally dependent on one another.

2   Nuclear Safety Regulations for Reactor Installations at Atomic Energy Stations 89/97 (NSR RI AS 89/97) defines nuclear safety as "harmonizing reactor installations and atomic energy stations with a DEFINED probability of avoiding the emergence nuclear accidents." The following questions arise:
   • Who defines this probability?
   • What is this defined probability?

3   NSR RI AS 89/97 includes a set of principles for creating safety systems. Two of these principles contain absurdities:
   • The independence principle – concerns improving the reliability of systems by functional and/or physical division of channels (elements) so that the failure of one channel (element) will not lead to the failure of another channel (element). As it were, definitions of independence should include conditions in which the failure of one element does not depend on the state in which another element is (working, failing, turned off). If this were the case, the independence principle here would not contradict accepted theories of probability and reliability.
   • the back-up principle – concerns improving systemic reliability by implementing structural redundancy with respect to minimizing necessary and *sufficient* (author's emphasis) defined systemic functions. The word "sufficient" is superfluous here, because if sufficient functions are the baseline standard, then why devote energy to substantive improvements? This ambiguity may lead to serious misunderstandings.

4   NSR RI AS 89/97 defines maximum reactivity as that which can be realized in a reactor during removal of all materials that impact reactivity and other extracted absorbers for the start of the removal process ... There were reactor accidents, when the absorber did change its geometric position (it was not extracted), while the absorber's effectiveness did change from the perspective of physical-mechanical processes.

Dr. Volkov has provided extensive commentary on all of the preceding observations. Definitional ambiguities, as a rule, lead to varying interpretations of documents, and consequently, to mistakes in technical safety compliance. As a result, this leads to increased negligence by state agencies with respect to the safe handling of nuclear materials. For example, qualitative risk evaluation procedures are well understood, but quantitative damage evaluation methods remain unclear. This uncertainty opens up possibilities for bureaucratic malfeasance. Upon examining the definition of the independence principle as it appears in official documents, it becomes clear that there is no requirement to review a particular event as being dependent on the failure of an element, when that other element has been removed from operation, for example, as a preventive measure. Such interpretations may lead to unnecessary expenditures on additional back-up equipment or, conversely, result in the lack of necessary back-up equipment out of cost-saving interests, which could lead to diminished safety levels.

Finally, a review of the Russian state agencies overseeing the safe handling of nuclear materials should be conducted. A starting point for such a review should be an understanding of the US system of nuclear non-proliferation policy. Figure 2.4 provides a model of the Russian system, based on the US system.

Creating such a system for relevant Russian state agencies remains a complicated task, as the process of restructuring relevant ministries and agencies was only recently completed. At the moment, Russian legislation relating to the handling of nuclear materials is transparent, but shortcomings still need to be addressed.

Conclusions and recommendations

This chapter has explored environmental issues related to Russia's nuclear industry in two main areas. First, attention focused on the basics of nuclear energy and nuclear waste production in order to lay groundwork for discussing mishandling of nuclear waste and problems of proper nuclear waste disposal. Next, the chapter examined norms and legal issues with respect to Russia's nuclear industry, noting an important legal change that may further affect the industry's impact on the environment.

In light of the previous discussion, the following factors have been identified as significantly impacting the effectiveness of environmental crimes due to the handling of nuclear materials:

Social factors

Most atomic energy stations and nuclear fuel production sites are the sole employers in the towns where they are located. Accordingly, closure of those facilities could lead to widespread unemployment. As a result, workers will continue to accept the only source of income in their region, despite continuing violations of production standards and subsequent damage to the environment and worker health.

As rule, highly qualified specialists work at atomic industry facilities. With proper financing from federal authorities, many of these specialists will be able to independently resolve issues relating to protecting the environment from radioactive waste.

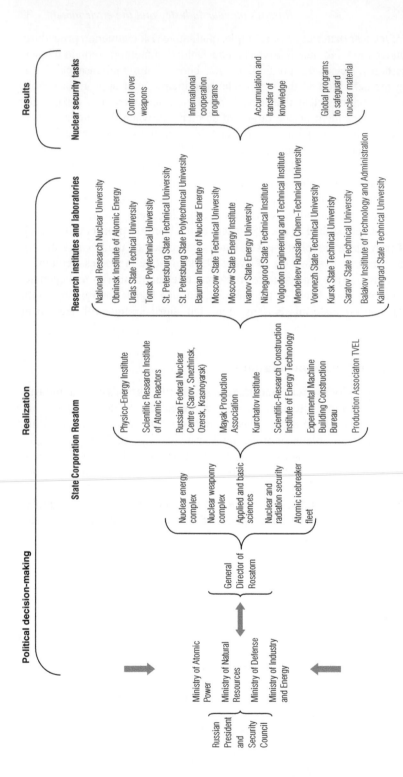

Figure 2.4 Structure of non-proliferation policies in the Russian Federation (based on materials provided by Mark Leek, Senior Research Scientist at the Pacific Northwest National Laboratory and Co-Director of the Institute for Global and Regional Security Studies at the University of Washington; diagram reworked by Dmitry Samokhin and Alexander Nakhabov, Obninsk Institute for Nuclear Power Engineering)

Technical factors

In addition to improving laws and regulations, ensuring that nuclear facilities have modern equipment and qualified staff remains vital. When a facility lacks the right equipment and qualified staff, it cannot adhere to standards and regulations. Barriers to fixing these problems include poor financial management and inability to replace older equipment. In other words, technical improvements to existing nuclear facilities alone will not solve all problems. At a minimum, personnel working at facilities with updated equipment must complete special courses to increase their professional qualifications. Ideally, efforts would be generally toward preparing a new generation of specialists, who initially gain experience working at higher education facilities, to replace the older generation of personnel at nuclear fuel sites. At the moment, the reality of an aging workforce in institutes charged with preparing specialists for work in Russia's nuclear industry poses numerous challenges. Unfortunately, talk about the necessity to retain and transfer knowledge in the nuclear industry, one of Russia's leading sectors, has thus far remained just that: talk.

Organizational factors

At the moment, Russian laws do not adequately specify procedures for managing nuclear facilities after they have been modernized, nor do they outline the conditions under which modernization should take place. Although there are persuasive arguments for and against permitting environmental specialists into nuclear facilities, this question should be addressed separately in the near future.

Shortage of qualified individuals (managers, environmentalists, nuclear technology specialists) who can guarantee safe handling of nuclear materials are troubling. Delays in improvements to existing facilities depend a great deal upon personnel qualification and training. High-quality work performance has much to do with the existence of training centers and special courses at nuclear fuel cycle facilities, where workers can improve their skills and gain experience with different operations. For personnel already working at atomic power stations, the schedule for passing specialized training is clear. However, the training and preparation schedule for personnel at other nuclear facilities, such as academic research centers, is not organized as well as it could be.

Returning to issues related to modifications of article 50 of the Russian Federation law "On protection of the environment," which allows for foreign spent nuclear fuel to be brought to Russia for processing, it is worth reiterating that Russia receives $20 billion for this service. A legislative project exists that would divert $3.5 billion to the national budget, $2.5 billion toward updating nuclear energy facilities, $7 billion for new technologies, storage facilities, and reactors, and $7 billion for special environmental programs. Obviously, this law should be passed and strict control over distribution of funds should be implemented. However, it is obvious that there is no guarantee that this money will actually be used in the way it supposed to be.

Legal factors

Before adopting laws that prohibit certain types of activities at nuclear facilities, it is necessary to prepare the industry for a "seamless" shift to a new legislative context. The worst-case scenario will see nuclear facility workers forced to work by old methods and on old equipment, in spite of new laws. In such an instance, sanctions would not be imposed because the nuclear facility is working on a government contract. As a result, a contradictory scenario appears: on one hand there is a law, but on the other hand the short-term goals of the government mean that the law can be disregarded.

Some Russian laws are in part aimed at populist goals, such as "keeping up with human progress." Fellow G8 members are satisfied, but certain laws have little hope of actually being implemented. In addition, insufficient measures are being undertaken to improve intra-industry cooperation (with Rosatom and with the naval fleet). Procedures for developing and approving documentation are drawn-out and complicated. Furthermore, financial difficulties make it difficult for companies to comply with laws and regulations in a timely fashion.

Nevertheless, it must be noted that the situation in general has improved greatly since the 1990s. In addition to increased government outlays, Russia's nuclear industry has been financed by foreign governments and international organizations. Laws are being changed and modernization of technical equipment continues. However, the fundamental shortcoming remains the slow pace of change, largely due to the factors noted above. As a final note, the media has failed to take a balanced view of environmental problems associated with the nuclear industry. Certainly, a more informed and constructive public would be a key ingredient to successfully addressing this important issue.

Notes

1   Federal Law No. 170-FZ RF "On the use of atomic energy," 21 November 1995.
2   P. W. Beck, "Nuclear energy in the 21st century: examination of a contentious subject," *Annual Review of Energy and the Environment* 24, 1999, 114.
3   U. Kroish, V. Noimann, D. Appel, and P. Dil, *Publications Dedicated to Nuclear Questions No. 3: The Nuclear Fuel Cycle*, Berlin: Heinrich Böll Foundation, 2006.
4   Ibid.
5   Ibid.
6   Ibid.
7   Ibid.
8   Ibid.
9   Ibid.
10 Y. Marignac and X. Coeytaux, *The Unbearable Risk – Proliferation, Terrorist Threats and the Plutonium Industry*, report prepared for The Greens/European Free Alliance in the European Parliament, June 2003.
11 K. Olsen, T. Farmer, M. Thomas, and A. Mitroshkova, Open World Center Presentation, Richland, WA, 2009.
12 V. Larin, *Russian Atomic Sharks*, Moscow: KMK, 2005, 380.
13 Ibid.
14 Ibid.

15 Ibid.
16 A. Agapod, R. Artiunian, S. Brykin, S. Kazakov, and G. Novikov, "The Problems of Radioactive Waste and Spent Nuclear Fuel: Perspectives on Solutions," *Atomnaia Strategiia*, September 2004.
17 Y. Vishnevskii, V. Iryushkin, A. Kislov, et al, "On security regulations for handling liquid radioactive waste accumulated in storage facilities at PO 'Mayak,' the Siberian Chemical Combine, and the Mining Chemical Combine," *Gosatomnadzor Rossiiskoi federatsii, Biulleten* 3, 2000.
18 Ibid.
19 Ibid.
20 Ibid.
21 A. Altunin, "Optimization of multi-level hierarchal systems based on fuzzy set theory and self-organization methods," in *Problems of Tyumen oil and gas*, A. Altunin (ed.),Tyumen: Tyumen State University Press, 1986, 68-72; A. Altunin, "Models and algorithms for decisionmaking in ambiguous situations," *Monographs*, A. Altunin and M. Semukhin, (eds), Tyumen: Tyumen State University Press, 2000, p. 352; V. Borshchevich and V. Botnar *Fuzzy Modeling and Problems of Interpretation*, Chisinau: KPI, 1984, 13.
22 M. Lizikova, "International environmental and legal aspects of nuclear security and energy charter," doctoral dissertation, Moscow, 2005, 174.
23 R. Gabdullina, "General characteristics of international legal norms on liability for crimes related to nuclear materials," *Chelyabinsk Universitet, Biulleten* 9/5, 2003, 136.
24 Materials Environmental Reviews Under the National Environmental Policy Act (NEPA), 2011. The U.S. Nuclear Regulatory Commission, http://www.nrc.gov/materials/active-nepa-reviews.html (accessed 10 May 2011).
25 *Developing Environmental Protection Policies, Programs and Procedures at Class I Nuclear Facilities and Uranium Mines and Mills*, Regulatory Guide G–296, Ottawa: Canadian Nuclear Safety Commission, 2006; *Environmental Protection Policies, Programs and Procedures at Class I Nuclear Facilities and Uranium Mines and Mills*, Regulatory Standard S–296, Ottawa: Canadian Nuclear Safety Commission, 2006; L. Keen, "Regulatory controls for radioactive sources and nuclear materials: a Canadian perspective," IAEA General Conference Senior Regulators Meeting, 2004; *Protection of the Environment*. Regulatory policy P-223, Canadian Nuclear Safety Commission, 2001.
26 S. Klem, "Environmental crime and the role of ICPO-Interpol," Proceedings of the Third International Conference On Environmental Enforcement, Oaxaca, Mexico, 1994, 335–341.
27 EU and Russia Cooperation Program, Harmonization of Environmental Standards II. *Final Technical Report: Environmental Liability and Insurance*, 2009.
28 Ibid.
29 "Fundamental aspects of Rosatom's environmental policy," 2010, *Energoinform*, http://www.energoinform.org/normatives/rosatomecopolicy.aspx (accessed 20 May 2011).
30 M. Lizikova, 2005, *International Environmental and Legal Aspects of Nuclear Security and Energy Charter*, Moscow, 174.

3 Forest auctions in Russia

How anti-corruption laws facilitate the development of corrupt practices

Svetlana Tulaeva

In a state where corruption abounds, laws must be very numerous.
Gaius Cornelius Tacitus

Most scholarly studies illustrate that the transition to a market-oriented economy in Russia was characterized by the growth of informal and corrupt relations when the absence of effective market structures permitted alternative codes of informal norms to develop. The Russian forest industry sector is a perfect example of this.[1] In the Soviet era the timber-processing complex was part of the planned economy in which enterprises belonged to the state and gained unlimited access to forest resources. All enterprises were tied directly to the command-administrative economic system. Stockpiled timber was divided between timber-processing facilities. Part of the finished timber was for domestic sale; part was exported.

The demise of the planned economy led to a situation where former ties between enterprises disintegrated and they lost their former buyers and suppliers. An additional hardship for many companies was their on-going social responsibility to the local population. The majority of timber-processing enterprises were located in towns and settlements. Therefore, they assumed the basic expenditures for supporting the infrastructure of the towns. When a planned economy and government subsidies were in place, a system of social responsibility was possible. However, without subsidies, enterprises were not able to support community projects that did not bring in profits. As a result, an array of bankruptcies of small timber-processing enterprises ensued in the 1990s and they were not able to underwrite community establishments to which they were tied in the Soviet era. This led to the twofold reduction of timber-processing industries in the 1990s and pushed small villages to the brink of extinction.

In order to support themselves during the transition period 1987–1995, enterprises actively used informal networks and negotiations; methods that were holdovers from the Soviet era. Many researchers characterized those methods as "virtual" because they were based on illusions about all of the key economic indicators: prices, trade, taxes, and budgets. The core of the illusions was the assertion that the economy was much more powerful than it actually was.

Such an economy can function only when market competition is substituted with command-administrative decisions.[2] The survival of some enterprises was supported not by their effective economic activity, but by their informal ties to administration representatives. Academics called these enterprises "economic zombies."[3] Clashes between these two systems, command-administrative and market, revealed that methods from the Soviet epoch had not disappeared completely; rather, they mutated into something new.

In the course of two decades, reform of the Russian Forest legislation should have created the environment necessary for developing Russian timber business on a market basis and liquidating illegal and corrupt practices involving the misuse of forests. However, this was not so. Constant changes in the laws, poor control over the laws' implementation, the absence of transparency and accountability in adopting decisions by government agencies did not augur well for the formation of a stable (institutional) environment.

Methodology

This research is grounded on an empirical approach to the study of law. Law is considered to be the result of a complex interrelationship among various interest groups in a dynamic political and social environment.[4] The research focuses not on the formalized "letter of the law," but instead on the practical implementation of laws by various actors. This approach negates the invocation of classical notions of legal or illegal; instead, it considers legality as a continuum between the poles of complete legality and complete illegality. Along this continuum exists a multitude of practical applications of law, which fluctuate according to their degree of correlation to the legal construct. Practices may be legal in form, but deform the spirit of the law; conversely, practices may be formally illegal but may conform to the actors' accepted notion of justice.[5] Cohn defines this situation as "fuzzy legality." He depicts several scenarios that develop this notion: an insufficiency of laws, unequal distribution of power and/or authority, selective enforcement, the presence of parallel and conflicting mechanisms of legal control and action, and the absence of control over the enforcement of laws.[6]

A situation of fuzzy legality arises in the absence of a solitary law that contains detailed and decisive instructions on its implementation, and/or when a law is ambiguous and open to interpretation. This might result in the manipulation of the law or the development of corrupt practices, resulting in the transformation of the law to something other than its originally intended purpose. This deflects attention from the law itself onto the actors involved in its implementation and, particularly, to the environment that enables utilization of the law in one or another unintended ways. Edelman and his colleagues demonstrate how the spirit of the law on discrimination in the workplace was transformed: the law stipulated that an employer must create procedures for workers to express their grievances of discrimination on the basis of race, gender, or other such characteristics. This stipulation was intended as a measure of protection for workers, but in practice it became a method whereby employers diverted workers' complaints away from lawsuits and did not lead to better conditions in the workplace.[7] A similar "flip-flop law" is described in an article by Volkov concerning the implementation of a law on commercial bankruptcy in Russia in the 1990s. The law, which was intended to lighten the burden on Russian enterprises of transitioning from a planned economy to a market economy, was instead utilized for hostile takeovers of companies by their competitors. The author shows that the Russian context of a closed structure of property rights and a prevalent shadow economy, combined with the incomplete nature of the law itself (a low threshold of liability and weak control over the actions of management), created the conditions for manipulating a law for purposes other than those envisioned.[8]

This research will focus on the manifestations of corruption arising from the implementation of a law on forestry auctions in Russia. As a rule, researchers cite the lack of transparency in Russian legislation as one of the fundamental reasons for corruption in Russian society. This might be attributed to the insufficiency and volatility of legislative acts, the high level of discretion of decision-makers, and the possibility of collusion in judicial processes.[9] This explanation remains insufficient, however, as it is focused solely on external conditions and pays no attention to the internal logic of these corrupt exchanges allowing the actors to "become a part of the public landscape."[10] This research uses an anthropological approach to studying corruption. Viewing corruption from the perspective of the corrupt actors the author explores the methods by which corruption becomes entrenched in everyday life of a society.[11]

In this chapter, the author strives to 1) show how the introduction of new anti-corruption laws provokes the development of new corrupt practices and 2) explain this phenomenon from an anthropological perspective.

The work is based on qualitative research methods:[12] analysis of documents concerning the management of forestry auctions and competitions, as well as semi-structured interviews with their participants. Materials collected in Leningrad oblast' in 2010 and in the Republic of Komi in 2006 provide the main empirical base. Interviews were conducted with various actors in the process, including:

*   representatives of large and small businesses that participated in forestry competitions and auctions;
*   directors of forest ranger stations who participated in preparing for and conducting the competitions and auctions;
*   representatives of regional administrations who were members of the competition commissions;
*   representatives of the auction commission of the Committee on Environmental Resources of the Leningrad oblast'; and
*   experts from environmental NGOs.

Documents utilized in this research can be divided into the following groups:

*   legislative documents of the Russian Federation;

- protocols of forestry auctions;
- reports of the Chamber of Accounts of the Russian Federation that contain evaluations of the effectiveness of Russian Federation laws;
- mass-media publications that discuss the law on forestry auctions.

Data received were coded and analyzed using the progressive approximation method [13] The use of various sources that reflected the perspective of all participants in the process and belonging to different time periods allowed for:

- triangulation of data;[14]
- examination of the phenomenon from various perspectives;[15] and
- the tracing of the trajectory of its development.

As a whole, the utilization of qualitative methods allowed for the identification of latent mechanisms used to transform the law in practice.

Forestry competitions and auctions in Russia, 1997–2006

The Forest Code of 1997 established forestry competitions as the fundamental mechanism for offering tracts of forest for lease. There are several key steps in conducting competitions: 1) determining the tracts of forest land that will be made available for lease in the competition; 2) preparing the documents and advertisements; 3) evaluating participants and identifying winners; and 4) creating a lease agreement based on the competition results.

All necessary organizational tasks related to the forestry competition were carried out by the municipalities' forestry management agencies (hereafter *leskhozy*). In order to conduct a competition, a commission would have been established, comprised of representatives of the oblast' and regional administrations, the *leskhoz*, and government environmental protection agencies. Competitions could have been open or closed, as determined by the chair of the competition commission.[16]

Information concerning open competitions was disseminated through mass media, and closed competitions were advertised only among potential participants. All information materials concerning the forestry competition were to have been produced no less than 30 days prior to the competition. Criteria used to evaluate contestants included: the amount of rent offered, the productive capacity to utilize the timber, length of time the enterprise had worked in the subject territory, prior forest revitalization efforts, potential job creation of the project, and the enterprise's participation in programs to benefit the subject territory's society.

Commissions were empowered to add criteria at their discretion.[17] In other words, the price offered was merely one criterion used to determine competition winners and was given the same weight as experience and corporate social responsibility. The length of the term of the lease was also determined by commission members, but was mandated to be between one and five years. The entity that won the competition would sign a lease agreement stipulating the land boundaries, the type of forestry utilization, the rental price and rent-collection terms, the responsibilities of all parties related to the preservation of the land, and the terms of payment for conducting forestry management work.[18]

In addition to forest competitions, it was possible to obtain a forest tract for short-term use for a period of up to one year through a forest auctions process. These auctions related to the use of particular plots of forestry land. Forest auctions were organized by regional-level forestry administration agencies, and were of two types: verbal auctions and "mixed" auctions (a method of combining verbal bids and written bids). The auction organizer determined which type to employ.[19] In contract to forest competitions, which offered a set of criteria which participants must abide, forest auctions relied solely on price to determine winners. However, in several instances, particularly vast tracts of forest were sold off in closed forest auctions, the bidders of which were chosen and invited by auction committee members.[20] The starting bid was established by the competition commission. As a rule, the bid ought not to have been lower than the minimal tariff plus expected commercial expenses. The auction organizers, at the request of the buyers, must offer a viewing of the forest land parcels that will be available for auction.[21]

The procedures for conducting forest-use competitions and auctions were vulnerable to manipulation at various stages. Both government employees and businesspeople exploited the vulnerabilities in a number of instances, leading to corrupt agreements between sellers and buyers. First, announcements concerning upcoming competitions were published such that only the "necessary" players learned of them: an announcement was either posted in a publication with very limited readership, or the entire publication run was purchased by the "necessary people." Therefore, in order to receive information about a forthcoming competition or auction, a direct link to the regional administration was in order.

Second, businesspeople frequently were ill-informed about which tracts of forest were to be made available in a competition or an auction. As a result, companies obtained the use of forest areas that were a poor match for industrial use. For this reason, auction participants tried to learn specific information about which tracts of forest would be made available in advance, using their informal connections within government to do so:

> By law the *leskhoz* must provide the plots of forest to companies at no charge, but in order to pad their budgets, they established a mechanism of payment for information related to the plots of forest. Money received for this information was spent by the *leskhoz* on forest management.[22]

Third, forestry competitions allowed for the shortlisting of participants using various criteria that frequently were difficult to evaluate objectively. This also created opportunities for manipulation. Often, in order to receive the desired tract of forest, companies were expected to have informal negotiations with representatives of the regional administration. And because the basic requirements of members of the auction commissions, other than to receive rental payments for the forestry tracts, were to deliver social assistance and provide jobs to the region, the informal negotiations included discussions concerning the contender's ability to provide social assistance to the region in the event of a win:

> And my support was very good. The head of our region and I visited the head of 'N' region and concluded an agreement on social partnership whereby we would provide the administration 450,000 rubles per year, in addition to the forestry tribute. How they utilize the funds is not my business. So we came to an agreement and when the competition was held we had tremendous support and nine of the commission members voted for us.[23]

Or:

> When we won the competition in 2003 the conditions were favorable. Consider first the solvency of the company, so that there were no debts for the forestry lease. Then there was financial support to the regional administration. The company accepted some responsibilities. The administration itself put these conditions on us – to do this and that.[24]

Often there was no accounting for how the funds were spent. Although large enterprises were able to direct their payments toward targeted projects, small companies had no idea how their social-service payments were allocated.

Sometimes the procedure of leasing forestry tracts allowed the administration to support small companies that were unprofitable, but that provided the sole source of jobs for entire villages:

> We never had big problems getting (forestry) leases, because our forestry company has been in business since '55. It has a name and experience. We produce electricity and wood. Everyone understands that if we shut down, two villages might collapse immediately.[25]

Further, in a number of instances entrepreneurs made attempts to conspire with one another prior to a competition or auction:

> It was very difficult, with a big fight … The competition was in 2004. I went to the Forestry Committee of the Republic of Komi, where they explained that the deal wasn't out there only for us. I say, well everything's understood, then I went to the other guy [a competitor – editor's note]. He and I get together regularly, and ride together to various events. And honestly speaking, we even went out drinking wine together once. In other words, we have a decent relationship. And he says to me, if it were your personal enterprise I wouldn't even compete against you. But it's NNN and I'm ambivalent toward them and I need forest. I say, no worries, I get it."[26]

Therefore, on one hand such informal relationships among businessmen and administrations concerning forestry leasing both softened the transition of the

Russian forestry resources industry from a planned-economy model to a market-economy model and allowed for the survival of neglected forestry-dependent villages. On the other hand, however, they encouraged corrupt practices. The practice of conducting forest competitions and auctions eventually became nothing more than seemingly legal on the surface but completely warped from its intended purpose. The law, initially intended to create competition and find a means of identifying the most effective user of forestry resources, ended up fostering insider deals between businesspeople and commission members.

In 2004 and 2005, the Ministry of Natural Resources amended the procedures for conducting forestry competitions and auctions with the intended goal of further defining the procedures and eliminating the existing opportunities for corrupt practices. The practice of conducting closed competitions and auctions was eliminated, as they allowed for lack of transparency in the distribution of forestry resources. Announcements concerning upcoming forestry auctions and competitions were now required to be published in official publications of the Russian Federation and its subjects. The leasing term of forestry tracts was increased, first to 49 years, then to 99 years.[27] The lessee was prohibited from reselling the forestry tract purchased at a forestry auction.[28]

Despite these amendments, the practice of conducting competitions and auctions was hardly transformed. As research interviews with entrepreneurs demonstrate, companies continued to rely actively on informal negotiations with administrations to obtain forestry tracts. Moreover, these corrupt practices intensified during 2006 in response to changes contained in drafts of the new Forest Code.[29] Because the new code would transfer ostensibly responsibility of forestry leasing decisions from the regional level to the oblast' level and would thereby complicate the process, companies began scrambling to obtain the forestry resources they desired. They utilized informal connections and negotiations with regional administrations in order to acquire long-term leases of forestry tracts that they deemed most economically valuable.[30]

Forestry auctions in Russia, 2007–2010

The drafting of the new 2006 Forest Code was accompanied by intense debate concerning improving mechanisms for leasing forestry tracts, which reflected fundamental tenets in worldwide discussions on the effectiveness of auctions as a means of distributing natural resources.[31] Factors that determine the effectiveness of auctions include:

- the level of discretion of officials making competition decisions: the greater the discretion of the auction organizer in determining selection criteria for winners, the greater the potential for manipulation of the competition;
- the presence of barriers to entering the auction: this decreases the number of competitors, which in turn creates a precondition for insider deals and does not allow for the establishment of true market prices;
- the level of transparency of auction procedures, including the presence of formalized and clearly outlined selection criteria, as well as accessibility of

information concerning upcoming auctions, implementation of auctions, and auction results;

- the format for conducting an auction (open verbal bids, closed bids, silent bids, mixed bids): research indicates that the method most impervious to insider deals is the silent written bidding process;
- the lease term: short lease terms suggest the possibility of frequent transfer of usage rights and do not foster sustainable resource use; furthermore, long-term leases allow for companies to overcome disagreements stemming from the paradox of government ownership of the land and private control of its use.

Russian legislators initially reviewed several basic variations of forest transfer: competitions, auctions, and a mixed transfer system. The Federal Anti-Monopoly Service actively promoted auctions, as they were viewed as an effective mechanism against monopoly formation and unfair competition in the sphere of natural resource exploitation.[32]

Big business likewise supported the auction system:

> We understand very well that auctions are more costly. If we want to talk about purely egoistic interests of companies, then cheaper is better. However, from the standpoint of a balance of interests – the interests of the government, society and the forest… then in my view the optimal form… that is, in considering internal beliefs, group interests, or some other relationships that have arisen, you cannot outwit this process and let one person in and not let another person in out of your own hidden self-interests.[33]

Medium and small businesses were the groups most interested in preserving the competition system, as it allowed for obtaining forest leases not on the principle of the highest price offered, but with deference to the societal contributions and worth of the enterprise to the local community: "We will not survive under the auction process."[34] Competitions and the mixed-transfer system were eventually eliminated as options, as they were considered more corruptible systems because forestry administrations and relevant local- and regional-level government agencies were the ultimate decision-makers. Auctions therefore became the default mechanism for transferring forestry via lease.

The next stage of discourse was to determine procedures for conducting auctions. Representatives of big forestry enterprises demanded that auction participants had to be companies with the capacity to utilize the forest tracts themselves. This essentially would amount to prequalification of participants using a set of nationally-standardized criteria.[35] However, this position was met with contention centering on the fact that intermediary ("middleman") services are not illegal, and therefore restricting access of participants to auctions would be improper.[36]

As a result, under the new Forest Code, the sole mechanism for acquiring a lease on forestry tracts became auctions, with unlimited access by all interested parties and the winners of which was determined solely on the basis of maximal rental price offered. Legislators believed that this guaranteed a transparent method of leasing forestry resources at the maximum profit to the government: "Ratification of the new Forest Code will provide entrepreneurs with equal access to forestry resources and will help establish modern competition in this sphere."[37]

In accordance with the new Forest Code of the Russian Federation, auctions for forestry leases are organized by the oblast'-level natural resource committees. Forestry leases are made available for two main goals: 1) the timber industry, and 2) the creation of recreational zones (community recreation areas, etc.). The lease term is 49 years. Auctions are open to all interested parties. In order to conduct an auction, the organizers must form a commission which includes representatives of the oblast'-level government and the *leskhoz*. In other words, representatives of regional administrations and *leskhozy* were excluded from the process.

The regional *leskhozy* retained only the function of preparing the technical documents for the auction. Announcements of an upcoming forestry auction must be published 60 days prior to the event in print periodicals predetermined by the oblast'-level government. Announcements must contain the name of the forestry area and land parcel number, land measurements, total parcel size, average yearly output of the area, and starting price per lot. Any entity or individual planning to participate in the auction must send the auction organizer an application for participation, which contains information about the applicant as well as documents confirming ability to pay a down-payment. Bids are submitted by participants either in sealed envelopes (a "closed form" of auction bidding) or are rendered openly during the verbal auction (an "open form" of auction bidding). The form of auction bidding is determined by the organizers. In the event of receipt of only one bid for a particular parcel, the contract is made for the reserve price (the "fallback price").[38]

The new Forest Code came into force on 1 January 2007. Nevertheless, experience has shown that the application of the new law in practice has not always achieved the goals set by the legislators. First, there have been shortcomings in the distribution of information. Despite the fact that announcements of upcoming auctions are all published in print periodicals accessible to everyone, the information provided to prospective bidders has proven at times to be insufficient for calculating the potential profitability and economic benefit of particular forest parcels:

> All the information on the parcel of land is there, except for the felling (harvesting) area – that information is to be determined later. And the information is from the 1994 forest survey. Much has changed since then and portions have been felled.[39]

Therefore, a prospective participant remains obliged to utilize informal connections and contacts to acquire more accurate information about the forest tracts being auctioned.

Second, the new law created conditions for the rise of new corrupt practices. On one hand, the new law eliminated lack of transparency in qualifying interested parties for participation and transferred the ultimate decision-making power from the regional level to the oblast' or republic level, thereby removing the possibility of insider deals between the regional administration and businesspeople. On the other hand, since it became more difficult to negotiate with members of the auction commissions, collusion among auction participants was further strengthened:

> All auctions are negotiated deals. We also participated in auctions. We found out the price ahead of time and divvied up our lots with the other participants.[40]

This is possible because the community of contenders vying for lease of the same type of forest tract is limited and the participants know one another. So the prospective contenders first discuss the distribution of forest with one another, and then submit official applications for specific lots. In some instances, lots acquired for the minimum price are then subleased to other members of the colluding group at a mutually-beneficial price.[41]

The lack of true competition in the majority of auctions is tangentially responsible for this practice of collusion.[42] For example, from January through November 2010 Leningrad oblast' conducted nine auctions to lease forest for timber harvesting and the creation of recreational spaces. One of these auctions was dedicated to creating a recreational zone and the remaining eight – at which a total of 71 lots were auctioned – were for timber harvesting. Of these, 53 lots were bid on by only one applicant. That is, 74.6 percent of forestry tracts auctioned in 2010 for felling timber were leased for the starting bid price. Thirty-five lots were auctioned for the creation of recreational space, of which 17, or 49 percent, were bid on by only one applicant, who therefore paid the starting bid price.

In 2009, there were five forest auctions, one for creating recreational spaces and four for the timber industry. Twenty-five lots were auctioned at the four timber auctions, of which 23 – or 92 percent – were bid on by only one applicant who paid the starting bid price. More than one applicant bid on each of the eight lots auctioned for recreational space.[43]

Experts working at non-governmental organizations (NGOs) continue to document violations of forest-leasing procedures. Foremost is the proffering of areas under particular environmental protection. For example, in Leningrad oblast' forest areas were leased for timber felling purposes on the territory of the Kurgalsk Nature Reserve (*zapovednik*),[44] and for the creation of recreational spaces on the Lebiazhie reserve.[45] According to the NGO Zelenyi Mir, under the guise of "pruning," one logging company felled more than 2000 hectares of healthy trees on the Kurgalsk Nature Reserve during 2006–2009. The forest land from that wildlife sanctuary was subsequently leased at auction to the logging company, and in 2010 was leased to another timber-industry company for more "pruning." Moreover, government land surveys indicate that the sanctuary shrank in size by more than 9,000 hectares from 2000 to 2007. (In 2000, the preserve's territory comprised 59,900 hectares, and in 2007 only 50,700 hectares.)[46] The government officials consider the apparent reduction of protected territory to be the result of mistakes in previous surveys.

In another instance, land belonging to the Lebiazhie reserve was leased for the creation of recreational areas. Furthermore, builders began to erect cottages on the leased land. NGO experts succeeded in bringing criminal charges against the lessees, among whom were representatives of the Leningrad oblast' administration. The Leningrad Inter-district Environmental Protection Procuracy brought a suit against the lessees and fined them for "removal of illegally-erected structures and recultivation of a destroyed soil layer."[47] In practice, however, nothing ended up being done.[48] This suggests that in the distribution of forest for lease, the government agencies remain a significant authority.

Audits conducted in 2007–2008 by the Chamber of Accounts of the Russian Federation and the Federal Anti-Monopoly Service identified a multitude of violations during the implementation of forest auctions throughout Russia. Violations in the procedures of conducting auctions included insider deals between purchaser and seller, restricting access to auctions to only legal entities, setting unduly low lease rates, and more.[49] Additional doubt was cast on the anti-corruption power of the new law at the end of 2007 with the uproar over auctions for the creation of recreational space in the Moscow suburbs. The Russian Forest Management Agency (*Roslezkhoz*) was the auction initiator, and the Moscow Forest Management Agency (*Moslezkhoz*) was the organizer. The bidding was conducted with widespread regulatory violations. For example, the period of accepting applications lasted only four hours, during which time gaining entry to the building where the bidding was taking place was made significantly more complex than usual, resulting in the inability for many prospective applicants to submit their applications. As a result, 990 hectares of land were leased either for nominal prices or with minimal increases of 5–10 percent on top of the starting bid. The price for leasing 100 metres of land for a period of 49 years was $25–500, whereas the market price at the time was between $10,000 and $15,000. Lessees included high-ranking public officials and prominent businessmen.[50]

This prompted legislators to outline more detailed instructions concerning the implementation of these auctions, which would prevent them from becoming spurious imitations of true auctions. The Ministry of Agriculture developed a tutorial guide in 2009, in accordance with which the starting bid could be no lower than the minimum lease price. Proving the ability to provide a down payment of between 10 and 100 percent of the starting bid for the auctioned item became a required condition of participation in forest auctions. Conditions governing the organization of an auction, such as the dissemination of information concerning upcoming auctions, and the acceptance of applications from prospective bidders, were further delineated. However, auction participants report that those amendments failed to change the application of the laws in practice:

> It's all the same as in the Forest Code. Only they complicated the procedure – more documents are required now, an official agreement to provide a down payment and other such items. The process itself remains as before, nothing has changed.[51]

At the time of this writing, the new law has been in effect for only two and a half years, but there have already been some notable outcomes from its usage. First, it has resulted in property redistribution and the exit of small-sized companies from the market. The latter cannot possibly contend with holding companies at an auction and cannot participate in informal negotiations, as they lack the requisite status and resources.

> Large companies come. Small lessees are pressured. This behavior comple-
> ments the rationale of the Code. It is, of course, beneficial to the oblasts. They
> are more governable, more transparent, more easily controlled.[52]

This process will intensify further in ten or twenty years when lease terms are up for current lessees.

Second, the new law does not increase the economic benefit to the government. Considering that in the majority of instances in Leningrad oblast' the forest parcels have been leased at the starting bid price, the efficacy of the auctions as a tool to increase government income is dubious.

Third, the new law does not foster environmentally and socially sustainable forest use, as it does not allow for the most efficient bidder to win. According to NGO experts' testimony, gross violation of environmental norms remains the same as before.[53] Furthermore, as discussed above, the law hastens the closure of timber businesses, which may be small, but which are essential to their communities as practically sole employers in forest-industry villages.

Finally, the new law has not eliminated corrupt practices in the timber industry as such, but instead has spurred their transformation. Whereas buyers and sellers previously engaged in insider deals, now prospective buyers engage in insider deals among themselves. In some instances, government officials continue to use their office and decision-making authority to obtain forest parcels of particular interest to them.

Corrupt behavior in the distribution and lease of forest tracts has significant consequences. First, it causes harm to the environment. Accounts of corrupt practices in forest competitions detail many instances whereby the implementation of forest competitions resulted in a lessee with the right government connections but lacking appropriate qualifications. That is, the company's ability to "negotiate" made it the competition winner, rather than its objective ability to carry out sustainable forestry usage. This issue is of particular concern with regard to territories under special environmental protection. For example, a logging company that has obtained a forestry tract with special usage restrictions could simply hire the "necessary expert" to define for it the terms of restricted use, and use this "expert determination" to its own benefit. A logging company's circumvention of formal regulations governing the use of leased forest, combined with poor monitoring on the part of the relevant government agencies, results in the deforestation of valuable lands and the destruction of key ecosystems.

Second, the local community may suffer from the loss of its forest area (recreational space, areas popular for mushroom-gathering and berry-picking, a source of deadwood necessary for home cooking and heating, etc.) For example, the ill-defined portion of the Forest Code concerning the lease of forestry parcels for the creation of recreational areas allows for the lease of those parcels for projects that in no way benefit the community. The Forest Code allows for the construction of temporary buildings on the parcels, which is manipulated for the construction of "temporary" vacation-cottage villages and spa/sanatorium areas. In a number of cases the forest parcels have subsequently been formally converted into the category of "agricultural lands," leading to the development of the forest parcels and denied access to the territory for residents, which is in direct violation of the Forest Code.[54]

Third, corruption in this sphere results in economic losses. Insider deals among prospective auction participants allow them to obtain forest at the starting bid price, which results in lower than optimal payments into the government budget.

Finally, corruption ensures the continuance and development of corrupt practices, in that it undermines formal laws and regulations and renders oversight worthless. In such a situation, informal agreements among participants necessarily become fundamental to business development. Similarly, failure to conduct informal agreements lowers the company's "competitive edge."

Auctions in other sectors of business in Russia

The forestry sector is frequently regarded as an arena of particularly intense development of informal and corrupt relationships. However, other sectors feel the negative effects of the application of auction laws as well. For example, the Federal Anti-Monopoly Service has made competitions and auctions a requirement of all government procurements and tenders. According to Federal Law 94, effective in 2006, all government purchases for sums higher than 100,000 rubles are required to be made according to "competitive procedures."[55] Tenders have become the chief method of carrying out competitive procedures. This is theoretically an equitable method of distributing orders for the delivery of goods, provision of services, and performance of contractual tasks for a cost and under terms agreed-upon in advance and published as such. A government contract is signed with the winner of a tender – a supplier who has submitted a proposal that meets the requirements of the competition and contains the most attractive conditions.

The goal of the legislators in creating this law was to combat corruption, create a competitive environment, and save money in the government's budget. However, the introduction of new anti-corruption regulations in the sphere of government procurement also led to unanticipated results in the application of the regulations.

First, the implementation of the law led to a decline in the quality of goods and services purchased. The tender process presupposed that the winning proposal would be that which carried the lowest cost, and did not consider quality to be an important evaluation criterion.

> For example, they held a competition for forest revitalization. A firm called *Roga i kopyta* (Antlers and Hooves) comes to bid. All the companies were prepared to do the work for 60,000 rubles. This firm says it can do it for

45,000. The rest, reputable forestry-stewardship companies, understand that this isn't possible. But we can't do anything about the situation. By law they won the competition. Time passed – of course they didn't do anything. But we have to report on this project. Then they come to us and say – well, give us a kickback and we'll abandon the project.[56]

Or:

We had an instance when we conducted an auction for janitorial services. An individual won the bid. We ask him, how are you planning to clean 4,000 square metres by yourself? We hunted him down for a long time. In the end, he refused to sign a contract. He said that someone had entered his bid in the auction for revenge.[57]

As in the case of forest auctions, legislators amended the law to reduce the risk of poor-quality work resulting from the tender process. However, the sole document that can annul a contract with an unscrupulous contractor or prevent an inferior bid from winning is the work order. The work order, however, cannot contain the trademark name of the good, and may contain only a description of its technical characteristics. Often the government official tasked with creating the work order is not well-versed in how to properly do so and safeguard against receiving inferior-quality, but low cost, bids.

Second, this law fostered the development of corrupt deals among prospective bidders, much as the auction law had in the forest sector.

We had scenarios where a person stands at the front door [of our building] and says: I'm bidding for such-and-such a price. Here's your 100,000 rubles. Now you don't bid.[58]

Third, the law gave rise to new types of insider deals between suppliers and buyers. For example, they could introduce additional conditions on prospective suppliers:

It is possible to narrow the circle of suppliers by requiring a certain delivery schedule. For example, let's say the tender was for some type of complication production, difficult software, or expensive medical equipment which could only be made-to-order. If you demand delivery within, say, three days, then only the one who has known about it in advance can deliver.[59]

It is worth mention that insider deals between suppliers and procurers sometimes occurred not only for personal gain but also to ensure that the procurement was of the appropriate quality:

Under the auctions system, quality suffers. There is always a danger that some devil is going to come out of a bottle and undercut everyone, or someone is going to use this for dumping. Therefore, relying on the scruples of suppliers will result in only losses.[60]

This danger encourages organizers of procurement tenders to use shadow methods of organizing competitions to avoid encountering swindlers.

In summary, as in the forest-auction experience, conducting procurement tenders did not eliminate informal and corrupt practices, but rather forced them to mutate.

Research results

The research followed the trajectory of development of the law on forest auctions. Legislators gradually introduced newer and newer amendments with the goal of matching the law as closely as possible to its purpose – the elimination of corrupt practices in the distribution of forest resources. Thus, the process of conducting forest auctions was defined with increasing detail: closed competitions were eliminated; responsibility for conducting them was transferred from one government agency to another; the lease term was extended; and procedures for announcing upcoming competitions were formalized and defined.

Beginning in 2007 auctions became the sole mechanism for obtaining forestry leases, and all auctions must be conducted by the natural resources committee at the oblast' or republic level. This eliminated the sculpting of selection criteria to favor one contender over others, which was previously commonplace practice by commission members, and ended insider deals between representatives of the regional administrations and businessmen.

Finally, in 2009 a guide to properly conducting forest auctions was developed. Despite these developments, the subject research demonstrates that the law created to quell certain corrupt practices ended up stimulating the development of a new set of corrupt practices. Moreover, these new corrupt practices allowed participants not only to attain self-serving goals, but also to achieve socially sound objectives (such as to support businesses that were unprofitable but vital to the economic security of a community, or to preserve the desired quality of a good obtained through the government procurement process.)

We can see that the "letter of the law" is not the only important component of an effective legal system. The environment in which the law is implemented and applied is vitally important.[61] In the Russian context, we can identify two key interconnected factors that govern the application of law. First, we have "fuzzy law" in Russia, arising from the fact that the legal system was formulated relatively recently and is characterized by a high degree of inconsistency and changeability. For example, the 1997 Forest Code underwent a revision in 2005, which was followed by the adoption of a completely new code which went into effect in 2007. This feverish pace of legal change strengthened the fuzziness of the existing judicial environment. Additionally during that time, both the structure of governance over forestry resources and the government agencies responsible for ensuring proper application of various laws were constantly in flux. That led to the absence of effective monitoring by government agencies of whether laws were being observed. Additionally, civil society lacked the capacity to monitor the observance of laws.

Second, Russian society has an inherent system of commonplace norms and values that allow for the inclusion of corrupt practices into everyday business. That system is respected and accepted on a par with formalized laws. In his study of corruption in African countries, de Sardan discusses the presence of various principles that promote the legitimization of corruption in society. Those legitimizing principles, as a rule, are closely intertwined and do not foster the development of pure forms of corruption. Instead, they give rise to specific types of behavior in accordance with which corrupt behaviors begin to be considered part of the societal norm.[62] Using information obtained during this research, we can identify the following types of legitimizing mechanisms applied:

The principle of "survival"[63]

The difficulties of the transition period left many logging communities on the brink of extinction, which gave rise to informal business practices as the single way to preserve certain enterprises and the villages dependent upon them.

> If it [the logging company – editor's note] leaves, the rest of us will face big problems. That is, the village will begin to die out. As long as the enterprise stays in business there, the village will survive.[64]

From this point of view, the violation of environmental regulations concerning the proper use of forest lands is not of overriding concern: "They preserve nature, but they don't care about us. What are we, dead wood?"[65]

The principle of "my inner circle"[66]

Interpersonal negotiations are considered a foundation for future formalized cooperation. Under conditions of high volatility and changeability, informal relations are utilized by participants as insurance against risk:

> Many people promote 'their service provider' not in order to obtain kickbacks, but because they can be certain that he will complete the work order well and on time.[67]

The principle of "counterattack"[68]

This presumes the growth of corrupt practices not as much for personal gain as for countering other corrupt individuals' aims.

> The more numerous the interested parties, the more leverage over the situation exists. If all the parties involved do not like us, they can say: pinch them, lie heavy on them till they pop.[69]

In other words, a company is forced to employ informal negotiations with other businessmen or government officials in order not to be squeezed out of the market. Unwillingness of a company's leadership to engage in the existing network of informal activities could cost the company leasing opportunities and therefore production capacity.

The principle of increased efficiency

This presupposes a situation wherein the circumvention of formal regulations improves material efficiency and time efficiency. "As an example, an enterprise violated logging transportation regulations, and in so doing destroyed the undergrowth. Regulations require the *lezkhoz* to levy a fine on the logging company amounting to three times the cost of reforestation, and send the fine payment into the federal budget. In practice, the logging company is not issued a fine by the *leskhoz*, but instead pays the cost of reforestation undertaken by it. This is beneficial to both the *leskhoz* [which receives direct payment for its reforestation obligation – translator's note], and the logging company."[70]

The tools for legitimizing corrupt practices described above bear two important consequences. First, practices of informal negotiations are considered an integral part of society, which underscores how routine corruption has become.[71] Second, they create an unwavering faith that without those practices nothing will get done. In this regard it is worth calling to mind Sedlenieks' comparison of faith in corruption with the Middle Ages era faith in the magi, whereby faith in one or another phenomenon becomes foremost in determining actions.[72] Faith in the ubiquity and effectiveness of corrupt actions gives rise to a certain style of behavior. The latter has a cyclical effect, ensuring the continuity of corruption.

Conclusion

Corruption is a more pervasive phenomenon than commonly thought. The discourse concerns not only misuse and abuse in the distribution of resources, but also an entire system of daily rituals and rights allowing this misuse and abuse to exist. The system transforms laws from being mechanisms to prohibit particular types of activity to being roadmaps of how to circumvent roadblocks.

Moreover, the research allows examination of the close ties between the level of corrupt practices, the degree of transparency in the decision-making process and the level of wellbeing of the environment. The exclusion of the public from the management of forestry resources, the circumvention of formal rules, and the misuse and abuse of office leads to the deforestation and destruction of protected areas, a low quality of forest regeneration, and more.

Despite the pessimistic review of anti-corruption legislation, it is essential to try and reduce the effects of corruption on the application of the existing laws. Additionally, it is important to employ various measures aimed to improve both the legislation and control over its application, as well as to alter the general environment. First, it is imperative to plug holes in legislation. The absence of certain conditions placed on prospective auction participants prohibits us from separating effective forest utilizers from unscrupulous ones. That is, the absence of minimum-level requirements concerning the companies' capacity to utilize and process timber allows for a case where an auction winner subleases his forest tract to another enterprise. Such requirements should be clear and concise and universally applicable.

Second, control over existing regulations must be enhanced. Government agencies neither monitor the implementation of any of the procedures nor investigate collusion among prospective lessees.

Finally, civil society must be called upon to monitor adherence to laws. This could include the participation of NGO experts at various stages of conducting auctions: provision of expertise during the preparation of tracts of forest to be auctioned, distribution of information about upcoming forest auctions, participation in auction proceedings, and provision of necessary monitoring over the activities of the lessee. Inclusion of NGO representatives in business-government relations would:

- bolster the transparency of transactions;
- create additional monitoring tools to observe the actions of both the companies and government officials;
- increase the information about the environmental and social importance of the forest tracts to be auctioned; and
- encourage and support inter-disciplinary dialogue in Russian society on a formalized level and not on the basis of personal negotiations.

Notes

1　M. Olsson, "Barriers to change? Understanding the institutional hurdles in the Russian forest sector," Lulea University of Technology, Sweden, 2004; I. Olimpieva, "The informal economy of forest use as a topic of research," *Neformal'naia ekonomika lesopol'zovaniia: uchastniki, praktiki, otnosheniia*, Moscow: Moscow Public Science Foundation, 2005.

2　C. Gaddy and B. Ickes, "Russia's virtual economy" *Foreign Affairs* 77, September/October, 1998, 5.

3　L. Carlson, N. Lundgren and M. Wilson, "If only money grew on trees: the Russian forest sector during the transition period" *Problemy teorii i praktiki upravleniia* 2, 2001; http://www.didaktekon.se/mats/pdf-files/RusChron.pdf (accessed 27 October 2010).

4　M. Cohn, "Fuzzy legality in regulations: The legislative mandate revisited," *Law and Policy* 23(4), 2001, 469–97.

5　Olimpieva, "The informal economy."

6　Cohn, "Fuzzy legality in regulations."

7　L. Edelman, C. Uggen and H. Erlander, "The endogeneity of legal regulation: Grievance procedures as rational myth," *The American Journal of Sociology* 105(2), September 1999, 406–454.

8　V. Volkov, "Hostile enterprise takeovers: Russia's economy in 1998–2002," *Central and East European Law Review* 23(4), 2004, 527–548.

9 M. Levin and G. Satarov, "Corruption and institutions in Russia," *European Journal of Political Economy* 16, 2000, 113–32.
10 J. P. Q. De Sardan, "The moral economics of corruption in Africa," *Journal of Modern African Studies* 37(1), March 1999, 25–52.
11 O. Pachenkov and I. Olimpieva, "Corruption, anticorruption and the position of the social science scholar," in *Fighting with Windmills? Socio-anthropological Approach to the Investigation of Corruption*, O. Pachenkov and I. Olimpieva (eds), St. Petersburg: Aletheia, 2007, 83–120.
12 S. Kvale, "The qualitative research interview," *Journal of Phenomenological Psychology* 14, 1983, 171–96.
13 L. Neuman, *Social Research Methods: Qualitative and Quantitative Approaches*, Boston: Allyn and Bacon, 1991.
14 Kvale, " Qualitative research interview."
15 C. Geertz, "Description: Toward an interpretive theory of culture," in C. Geertz, *The Interpretation of Culture*, New York: Basic Books, 1973, 3–30.
16 Article 34, Forest Code of the Russian Federation Regulations on the methods of conducting forest competitions for leasing forest sectors, Russian Federation, 30 September, 1997.
17 Ibid.
18 Ibid.
19 Regulations on the methods of conducting forest auctions, 11 August 1997.
20 Ibid.
21 Articles 43-45 Forest Code, RF, 1997.
22 Interview with a representative of a forest industrial enterprise, Republic of Komi, December 2006.
23 Interview with a manager of a forest industrial enterprise, Republic of Komi, December 2006.
24 Interview with a manager of a forest industrial enterprise, Leningrad oblast,' October 2010.
25 Ibid.
26 Interview with a representative of a forestry industry company, the Republic of Komi, December 2006.
27 In practice, forest tracts were never leased to a company for a period as long as 99 years.
28 Decree #97 by the Ministry of Natural Resources of the Russian Federation on 14 April 2005 in Moscow, "Confirmation of the Procedure for Organizing and Conducting Forest Auctions."
29 Interview with a representative of a forestry industry company, the Republic of Komi, October 2010. Interview with a representative of a forestry industry company, Leningrad oblast', October 2010.
30 Interview with a representative of a forestry industry company, the Republic of Komi, October 2010. Interview with a representative of a forest industry company, Leningrad oblast', October 2010.
31 A. V. Rodionov and A. A. Rogov, "Formalizing procedures for selecting foresters," *Issledovano v Rossii* electronic journal, 2004, www.zhurnal.ape.relarn.ru (accessed 27 October 2011).
32 Official site of the Federal Anti-Monopoly Service: http://www.fas.gov.ru (accessed 15 November 2010).
33 Interview with a manager of a forestry industry holding company, December 2006.
34 Interview with the director of a private enterprise in the Republic of Komi, December 2006.
35 Interview with a manager of a forestry industry holding company, December 2006.
36 See press reports. For example "A forest ranger and a lumberjack argued," http://www.industrydaily.ru/main-theme/20 (accessed 27 November 2010).

37 Official site of the Federal Anti-Monopoly Service, http://www.fas.gov.ru/fas-news/ fas-news_9316.html, (accessed 15 November 2010).
38 Articles 71–80 of the Forestry Code of the Russian Federation, 2006.
39 Interview with the director of a forestry area, Leningrad oblast', October 2010.
40 Interview with a representative of a small timber-processing company, Leningrad oblast', October 2010.
41 Interview with a representative of a timber industry company, Leningrad oblast', October 2010.
42 It is important to note that not all one-bidder auctions result from collusion, since a number of forest tracts that are auctioned are located in areas that are difficult to access and might be of interest to only one company.
43 Based on forest auction procedures for 2009–2010 published in the oblast' newspaper *Vesti*.
44 Russia is home to 85 strictly protected nature reserves (*zapovedniki*). Access to these reserves is normally restricted to scientists and those studying and caring for the flora and fauna.
45 The Kurgalsk and Lebiazhie Sanctuaries – wetlands of international significance. The Russian Federation accepted responsibility for their protection in 1994 in accordance with the Ramsar Convention of 1971.
46 Interview with Zelenyi Mir NGO representative, October 2010.
47 Decree of the Inter-district Environmental Protection Procuracy on the case concerning the Lebiazhie Nature Reserve; Zelenyi mir website, www.greenworld.org. ru/?q=lebiazhie2239. (accessed 10 December 2010).
48 Materials from the monitoring NGO, Zelenyi mir.
49 Official website of the Chamber of Accounts (*schetnaia palata*) of the Russian Federation; www.ach.gov.ru/ru/revision/reports-by-years (accessed 15 October 2010).
50 "Forest auctions have been discontinued," www.kvartirazamkad.ru. (accessed 14 December 2010); "Forest auctions in the outskirts of Moscow: Old mistakes and new dangers www.forest.ru (accessed 14 December 2010).
51 Interview with a member of the auction commission of the Leningrad oblast', October 2010.
52 Ibid.
53 Interview with an NGO expert working in Leningrad oblast', October 2010.
54 Interview with a member of the auction committee of the Leningrad oblast', 2010, article 41, Forest Code of RF, 2006.
55 Federal Law 94-F3 (F3-94) 21 June 2005.
56 Interview with the director of a forestry area, Leningrad oblast', October 2010. Discussion concerned work performed on forest tracts not leased to companies.
57 Interview with an organizer of government procurement tenders, October 2010.
58 Ibid.
59 Ibid.
60 Ibid.
61 V. Volkov, "On the other side of the legal system or why laws do not work as they should," *Neprikosnovennyi zapas* 42(4), 2005, http://magazines.russ.ru/nz/2005/42/ vv6.html (accessed 27 October 2010).
62 J. P.O. De Sardan, "The moral economies of corruption in Africa."
63 For a deeper analysis of this governing principle of behavior, see: J. Scott, *The Moral Economy of the Peasant: Rebellion and Subsistence in Southeast Asia*, New Haven: Yale University Press, 1976.
64 Interview with a representative of a regional administration, Republic of Komi, December 2006.
65 Interview with a resident of a forest industry village, Republic of Karelia, July 2009.

66 A similar set of principles concerning informal relationships is described by Hernando de Soto in, *The Other Path: The Invisible Revolution in the Third World*, New York: Harper and Row,1995.

67 This quote is from A. Boiarskii, "Under the hidden tender," www.4tender.ru (accessed 3 November 2010).

68 A similar set of principles concerning informal relationships is described by E. Paneiakh (University of Michigan) in the workshop paper *Rules of the Game for Russian Entrepreneur: Coping with Uncertain Normative Environments*, Workshop on Institutional Analysis September 16–21, 2006, Boulder, Colorado, USA, 2008.

69 Interview with a manager at a logging company, Republic of Komi, December 2006.

70 Interview with a manager of a forest industry holding company, December 2006.

71 For a more detailed discussion of baseline corruption see I. Olimpieva, "Baseline corruption in medium and small businesses: The weapons of the weak," in Pachenkov and Olimpieva, *Fighting with Windmills?* 213–31.

72 Sedlenieks, K., "Parallels between Latvia and Zande: Corruption as voodooism in Latvian society during the transitional period," in Pachenkov and Olimpieva, *Fighting with Windmills?* 191–212.

4 Combating corruption and organized crime in the forest sector of the Trans-Baikal Territory

Yekaterina Pisareva

The Russian Federation is home to enormous forest resources. The total forest area in Russia is 1182.6 million hectares, accounting for 22 percent of global forest cover. One quarter of the world's timber resources is contained within the territory of Russia.[1] Forests as a major natural resource have long been under state protection. The current state of forests in Russia, however, makes us wonder what we will leave to our descendants. Illegal logging, combined with permanent forest fires, presents a depressing picture. Statistics indicate that every year, 10 to 35 million cubic metres of wood are illegally logged in Russia. Moreover, the Eastern Siberia–Far East region accounts for about one-third of this amount (in 2007, it was seven million cubic metres).[2]

Unfortunately, many of us still have not learned to treat with care what nature gives us. This is the ethical side of the issue. However, illegal logging also harms national and regional economies. The Trans-Baikal Territory has vast forest resources, which, fortunately, are renewable, but only if those who cut and burn the "green gold," will restore those resources. The profits that the state uses to supplement the budget through the use of forest areas end up in the coffers of individual regions, including those of the citizens and various organizations dealing with the forestry business.

The transition to a market economy in Russia produced many positive results, but also gave rise to numerous environmental crimes. The uncontrolled development of market relations, the undervaluing of and lack of control over natural resources, freedom of entrepreneurship, inadequate legislation, and an attitude of consumerism in relation to the environment resulted in numerous violations of law regarding the use and protection of forests.

Annual losses from the illegal use of forest resources are in the millions of rubles (it is not possible to present the exact amount because of the high latency of illegal logging and limitations in the illegal logging damage evaluation regulations). The situation is further complicated by the fact that China – the largest importer of timber and timber products – is in close proximity to the Trans-Baikal. Recently the Chinese industry's demands for Siberian wood have grown exponentially.

During the investigation of criminal cases of illegal logging and timber theft, 72 units of motor transport and special vehicles and 358 chainsaws used to commit

Table 4.1 Criminal Code statutes relating to corruption in Russian Federation

		2009	2010
158	Theft	77	74
159	Fraud	1	1
160	Misappropriation and embezzlement	3	0
171	Illegal entrepreneurship	1	0
174	Legalization (laundering) of money or other property acquired by a person due to committing a crime	1	0
171.1	Production, acquisition, possession, transportation or sale of unmarked goods and products	14	0
188	Smuggling	19	7
285	Abuse of power	2	2
286	Abuse of office	2	0
290	Bribe taking	0	0
292	Forgery	6	1
293	Negligence	2	5
260	Illegal logging of forest plantations	889	950
	Total	1017	1040

the crimes were confiscated. In the court cases, 544 persons were convicted and 30 persons were sent to prison, 442 guns and 41 vehicles were confiscated by the state.[3]

In addition, the courts recovered more than 25.3 million rubles from the criminal offenders as a result of prosecutors' claims for reimbursement for the damage caused by illegal logging. More than 14 million rubles in damages was paid voluntarily. However, despite the increase in cases of illegal logging, the detection of such crimes has not improved.[4] A statement prepared by the Trans-Baikal Prosecutor's Office shows that 47.6 percent of crimes in 2010 were in the timber industry.[5]

Research methods

The methodological basis of the research was a sociological survey of law enforcement officers, investigators, and prosecutors at police and forest agencies, the inhabitants of the town of Chita and areas of the Trans-Baikal Territory. In all, 107 persons were surveyed, most of whom have direct contact with the problems of forest crimes in Trans-Baikal. A set of questions was also carefully designed for lengthy interviews and discussions with practitioners. Numerous materials on addressing corruption and organized crime in the forest sphere were also analyzed, including newspapers, journals, radio programs, and the internet.

The theme of our research is the study of organized crime and corruption in the forests of Trans-Baikal. Trade in forest products has very high liquidity and attracts significant capital. The wide circulation of money entices certain individuals and crime groups. Groups that control wood-processing plants receive a large percentage of the revenues of these firms. This is the payment for a so-called *krysha* ('roof' in English) that translates into a form of protection from competitors, including physical protection for the firm's activity, or the removal of obstacles to the business, especially from government agencies. In addition, a significant number of crimes covered by Article 260 of the Russian Federation's Criminal Code are committed by a group of people, groups of people by preliminary conspiracy, or an organized group of people. However, as police investigators point out, it is very difficult to prove the act of illegal logging by an organized group of people. It is much easier to provide proof in cases of timber smuggling, purchase of illegal timber, and sale of timber materials.

Corrupt connections of state officials and municipal authorities with woodsmen became a topic of discussion not long ago, when the fight against corruption at various levels became the top priority. At the national level, Presidential Decree of 13.04.2010 N 460 "On the National Anti-Corruption Strategy and the National Anti-Corruption Plan for 2010 – 2011" was adopted. The purpose of this anti-corruption strategy was to eradicate the causes and conditions leading to corruption in Russian society. To achieve the objectives of the National Anti-Corruption Strategy the following tasks should be undertaken in this order:

1   updating of legislative and institutional frameworks for combating corruption;
2   streamlining law enforcement and implementing administrative directions for combating corrupt behavior and ensuring the reduction of corruption;
3   promoting anti-corruption norms of behavior, including the use of appropriate enforcement measures in accordance with the laws of the Russian Federation.[6]

In the Trans-Baikal, activities against corruption and organized crime are gaining momentum. This is mainly because of tightening prosecutorial inspections of the timber and timber products trade. At present, special internal affairs units directly involved in the identification and detection of crime in the forest sector have been created, as well as units and officials in the prosecutor's office who supervise compliance with criminal and criminal procedural law.

This study aims to suggest improvements to legislation and to the practice of using standards of criminal liability for corruption crimes committed in the forest sector, as well as to develop recommendations for an effective response to organized crime activities. To achieve this goal we:

1   review normative-legal acts aimed at combating corruption and organized crime in the area of forest protection from criminal attacks in Russia and the Trans-Baikal Territory;
2   analyze statistical data on the number of violations and crimes in the forest sector in the Trans-Baikal Territory and the results of the investigation of such crimes to study the judicial practice;
3   conduct research and analyze data (questionnaires, interviews, analysis of publications in media). The research findings are also based on questionnaires,

interviews, and an analysis of the materials of criminal cases, a content analysis of media materials, an expert survey of law enforcement officers;
4   identify the presence or absence of organized crime in forest protection in the Trans-Baikal Territory;
5   identify the presence/absence of facts of organized criminal activity in the field of forestry;
6   evaluate the effectiveness of countermeasures to organized crime and /or corruption crime;
7   develop recommendations to improve measures for combating organized and/or corruption crimes and apply them in practice.

Countermeasures to organized crime in the Trans-Baikal forest sector

Article 260 of the Russian Federation Criminal Code provides liability for the illegal felling of forest timber by a group of persons (paragraph a, Part 2), a group of persons by previous concert or an organized group (Part 3). It is considered to be a preliminary conspiracy if several people take simultaneous and mutually agreed upon actions to achieve a common criminal purpose during the process of illegal logging. The perpetrators are persons who commit acts of separating the trunk from the root, that is, cutters, and those who have provided direct assistance to them and participated in the process of felling.

A crime committed by an organized group, which is provided in Article 260 of the Criminal Code, is a crime committed by a cohesive group of people who collaborate and plan to commit one or more offenses (Part 3, Art 35). Cohesiveness is defined as a high degree of organization, discipline, and adherence of the group to its organizational structure; the presence of unique, distinct forms and methods of criminal operations; and the groups' perseverance. An organized group can be created for commitment of one or several crimes and, in contrast to a group of persons by previous concert, co-conspirators have specific roles. As practice shows, not many people are convicted for a crime committed by an organized group under Art. 260 of the Criminal Code. According to our survey, 53 percent of respondents said they had heard about organized criminal acts from colleagues, reports, and the media, but have no personal knowledge of them. Data obtained from the Information Centre Trans-Baikal Regional International Affairs Office between 2009 and 2010 reveal that only two crimes committed by an organized group of persons described under Article 188 "Smuggling" have been registered.

An example of a case involving a forest crime by an organized group of individuals recently prosecuted is as follows: The deputy of the Legislative Assembly of Region "F", together with individual "P" organized timber smuggling worth over 60 million rubles. In addition, they committed tax evasion. As a result, the prosecutor recovered property damage in the form of unpaid taxes in favor of the state in the amount of 10,286,611 rubles 31 kopeks.[7] During the interview with Chita interdistrict environmental prosecutor Z. V. Diachkova, held 3 February

2011 in the Prosecutor's Office of the Trans-Baikal, we asked her a question about the extent of illegal logging by organized criminal groups in the Trans-Baikal. Diachkova replied:

> Five to six years ago organized groups often committed illegal felling. Chainsaws and other equipment were brought into forests under armed guard; criminals would quickly cut wood and transport it out of the forests undetected. Today this activity is seldom seen, not because the fight against forest crimes has become successful, but because, in fact, there is little valuable timber left to cut down. It is unfortunate that the best timber is stolen with only young forests remaining.

Indeed, based on the statistics for the last two years, cases of organized crime in illegal logging and timber sales in the Trans-Baikal have not been registered. The perpetrators of organized crimes in the forest sector, especially timber smuggling abroad, is extremely pprofitable and hard to detect.

A survey of the Investigative Division at the Chita district department of internal affairs (the investigation department directly dealing with illegal logging of forest plantations) also showed that illegal logging is most often committed by a group of persons by prior agreement. To make a formal accusation for committing illegal cutting of timber by an organized group of persons all features of an organized group, specified by criminal law, need to be proved, which is very difficult.

Furthermore, even if there is formal evidence of an organized group, in a court such a case may "fall apart" with a skillful defense lawyer in court. Therefore, in practice, prosecutors try not to take risks, but to institute criminal proceedings for actions actually committed, in order to punish all perpetrators. If one does not rely on proven facts of crime by organized criminal groups, but on characteristics of a criminal organization, it is possible to describe the crimes committed in the illegal logging and timber sale in the Trans-Baikal as manifestations of organized crime. It is impossible to procure and to sell illegal timber without the help of certain agencies, officials, and commercial organizations engaged in the forest business. As a rule, illegally harvested wood is transported, unloaded, and moved abroad by overcoming some obstacles overseen by the federal and regional laws to restrict and ban illegal business activities.

Often, those engaged in illegal logging are aware of the possible distribution channels for forest products and they work closely with entrepreneurs engaged in shipping and receiving timber. Law enforcement officials often refer to such phenomena as organized crime. To prevent illegal wood sales in the Trans-Baikal Territory, the law "On the organization of wood receiving and shipping stations in the Trans-Baikal Territory " was passed on 24 June 2009. This law provides for protection of the environment, the sustainable use of natural resources, and the prevention of illegal logging.

The law also defines the organization of the timber receiving and shipping stations and sets the rules for the reception, registration and transportation of wood in the Trans-Baikal Territory.[8] The law "On Administrative Offences" establishes liability for violation of the requirements of law of the Trans-Baikal Territory to the organization of the timber receiving and shipping centers in the Trans-Baikal Territory (Article 36).[9] However, despite the positive significance of this law, the penalty for the offense is unduly lenient. In addition, in order to prevent the illegal sales of wood, the practice of accompanying persons detained for possession of illegally logged timber to the place where it was sold was established. These measures help to prevent illegal business activities and assist in the prosecution of those who purchase or sell property known to be obtained by criminal means (Article 175 of the Criminal Code). As an example from judicial practice, individual "B" was caught on 29 October 2009 by the operational staff and officials from the "forest department" of the Interior Affairs Department of the Khilokskii district where the illegal logging occurred. Later, during the search operations at the wood receiving station in the village of Kharagun in Khilokskii district, a Chinese citizen bought the pine wood from "B," knowing that the latter had obtained it illegally. The magistrate judge found these persons guilty of committing crimes under Article 175 of the Criminal Code and fined them.[10] A conversation with law enforcement officials revealed that a network of well-honed business organizations actively using different ways to maximize the benefits from illegal logging and timber sales is operating in the Trans-Baikal Territory. This problem is especially acute in areas with good rail transportation where it is easy to quickly transport and unload the timber. A large amount of timber and timber products, including round wood, is not always correctly and accurately identified and recorded in the documents, as required by the law "On the organization of the wood receiving and shipping stations in the Trans-Baikal Territory." Nonetheless, inspections conducted by prosecutors and raids undertaken by law enforcement officers help prevent such violations and assist with the rapid identification and punishment of citizens and entrepreneurs who violate the law.

One of the most effective measures for countering forest crimes is what prosecutors call "operational information" or intelligence regarding the existence of illegal logging firms and illegal timber receiving stations in the district. This includes wire-tapping, recording of telephone conversations, and seeking out individuals who have information about the crime. It should be noted that in the Trans-Baikal territory most crime in illicit timber sale is not by organized crime but by group crime. It should be emphasized that organized crime groups commit crimes listed in articles 188, 190, 191 of the Criminal Code.[11] Illegal logging of forests and timber theft often occur as a crime by a group of persons by previous concert. Many individuals engaged in illegal logging and timber theft are economically disadvantaged and have no other way to make a living. They lack wood to stoke their furnaces and thus "steal" wood to survive. Organized crime groups, in contrast, are led by high-income individuals who seek to increase profits by bribing officials to enable them to transport and sell the illegal wood abroad.

Corruption crimes related to illegal logging and timber sales

It is well known that the timber trade is a very profitable business as a result of the high demand for timber from the countries of the Asia-Pacific region. As pointed out by Petrov:

> Forest resources have several features that lend themselves to corrupt deals. First, it is not easy to organize the monitoring of timber. Second, the multiple uses of forest resources allow paid access for one type of resource and the use of other resources for free.[12]

Illegal logging, without a doubt, involves administrative and criminal responsibility but, unfortunately, it often affects people without the resources to hire a lawyer to protect themselves from the punishing sword of the Russian justice system. Those are the socially and financially vulnerable populations who live mostly in rural areas and for whom the forest is the only source of income. These loggers, called "termites" by the local people, cause the least damage to the economy and the environment of the region, because they chop wood in small batches – two or three trees.

The real disaster for the forest sector comes from large timber processing organizations and individual entrepreneurs engaged in shadow timber businesses. Many entrepreneurs engaged in harvesting and selling wood have worked in this sector for a long time and know each other well. They also know about the possible distribution channels of illegally harvested timber and the ways of interacting with officials responsible for the condition of forests and their protection from criminal attacks.

In jobs where one may receive a large income, the phenomenon of corruption – an incurable disease of our process of management and control – inevitably occurs. Many authors believe that corruption has always existed in Russia, albeit in various manifestations. Russia ranks as one of the top countries in the world with respect to corrupt officials. In recent years corruption has put down even deeper roots into the political and economic system of the Russian state.

As a result, the top political leadership of the country has taken a course to eradicate this shameful phenomenon. Thanks to the measures taken in the country, and in our region in particular, the number of recorded corruption crimes has increased.[13] This does not mean that there is an increase in number of bribes but rather that law enforcement agencies have begun to investigate such crimes more efficiently. Unfortunately, we could not obtain accurate statistics on bribe giving and receiving related to illicit timber trade from the officials in the Trans-Baikal Territory. There is only limited data on individual cases of bribery.

As stated in the federal law N 273-FZ, "On Combating Corruption," enacted 25 December 2008, corruption is:

> the abuse of office, bribery, bribe taking, abuse of power, commercial bribery or another unlawful use by an individual of his official position ... for the purposes of receiving advantages in the form of money, valuables or other

property or property-related services, other property rights for himself or for third parties, or unlawful provision of such benefits to the specified person by other individuals, or the commission of such acts on behalf of or for the benefit of a legal entity.[14]

On this basis, the following crimes can be classified as corruption crimes in the forest sector: bribery (Article 291 of the Criminal Code), bribe (Article 290) abuse of authority (Article 285), abuse of power (Article 286), and forgery (Article 293 of the Criminal Code).

We obtained information from the Trans-Baikal Territory's Prosecutor's Office on the number of crimes committed in the timber industry in the Trans-Baikal. In 2008, five criminal cases were initiated on the grounds of misuse and abuse of power by forest management staff and other officials who misappropriated timber to themselves and committed other forest crimes. In 2009, 17 cases were initiated and in 2010, 12 criminal cases (7 of which were cases of corruption).

The Trans-Baikal Territory law "On combating corruption in the Trans-Baikal," dated 25 July 2008, develops the basic provisions of federal law and in many respects simply duplicates them with some attention to the features of the Trans-Baikal. The Trans-Baikal law "On anticorruption legal acts and their drafts" of 24 December 2008, which provided the order for presenting legal acts for expert evaluation has expired.[15] On 11 August 2009, the Government of the Trans-Baikal adopted an order "On approving the list of corrupt positions of state civil services in the executive agencies of state power of the Trans-Baikal," which included the office of the Forest Service and its territorial departments directly involved in forest protection and in the control of timber trade.

The Trans-Baikal 26 September 2008 adoption of the law on the "Development of the timber industry in Trans-Baikal Territory (2009–2013)" is very positive and has as one of its goals the decrease in exports of unprocessed timber.[16] Unprocessed timber products, also known as "round wood" are exported in huge quantities outside the Russian Federation, particularly to China, where manufacturers then use our wood to build flooring and furniture and sell them to other countries. As a result, Russia gains no benefits and suffers from denuded forests.

In summary, we have investigated the regulatory-legal acts of the Russian Federation and the Trans-Baikal Territory aimed at curbing corruption and the related crimes. Unfortunately, there is not a single regulatory act that contains instructions for curbing corruption in the forestry sector. It is extrememly difficult to adopt several legislative acts, especially at the regional level, on specific issues or sectors of life that relate to the fight against corruption. However, certain provisions of laws and other acts provide guidance on the need of combating corruption in our society. Such acts include instructions and orders for prosecutors in offices of the Trans-Baikal Territory who have the authority to supervise of anti-corruption measures at different levels.

We conducted an anonymous survey of individuals, employed in law enforcement for more than six years. Their replies illustrate that 52 percent of respondents are aware of corruption cases in the forest sector. Figure 4.1 shows

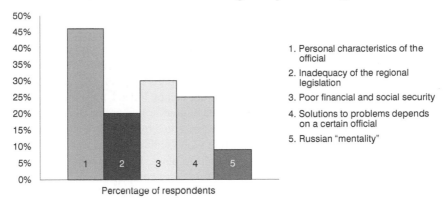

Figure 4.1 Leading causes of corruption in the timber trade of the Trans-Baikal Territory: opinions of law enforcement officials with more than six years of experience

which phenomena the respondents believe are the main causes of corruption in the timber trade in the Trans-Baikal.

Figure 4.1 also shows that the official's personality traits are the main cause of corruption, a subjective factor that ranks in the first place. This means that special scrutiny should be given to the personal characteristics of candidates seeking positions that involve the distribution of natural resources. Individuals with insatiable thirsts for profit would ideally be denied official positions in law enforcement that involve controlling the timber trade and conserving natural resources. In such cases, a constant rotation of personnel may help to minimize the risk of creating so-called "feeders," which sometimes occurs when one person holds the same position for a long time, becomes well informed in all the nuances of the timber harvesting, and becomes well known to entrepreneurs and persons engaged in illegal timber harvesting. However, given the peculiarities of Russian mentality it seems inevitable that anyone with a low level of legal consciousness and high material needs sooner or later will start to cooperate with "black woodsmen." Therefore, it is imperative to raise the level of legal consciousness of the administrative personnel, law enforcement and regulatory authorities, and to improve their salaries and benefits, especially those with a high degree of responsibility.

Analysis of the local media demonstrates that corruption in timber procurement and sales in the Trans-Baikal Territory was discussed many years ago. For example, in 2006–2007 local newspapers published reports on bribe-taking by forest management and law enforcement officials in connection with their efforts to identify and combat illegal logging and timber theft.[17] According to the "Extra-Media" Trans-Baikal Information Agency, a local logger who requested anonymity said that the Trans-Baikal timber industry is rife with corruption. According to him, the entrepreneurs who are trying to work honestly and transparently cannot export timber for months, while some corrupt forest market representatives operate freely.

In the field of timber harvesting, widespread corruption crimes are related to the issuance of timber declarations, the examination of logging areas, the conclusion of lease agreements, purchase of contracts for forest stands, inspections, and raids to detect illegal logging and transportation of illegally cut timber. According to 89.5 percent of surveyed law enforcement officers, the corruption offenses are most often initiated by citizens and individual entrepreneurs. Indeed, there are not many extortion cases by regional state forest management or law enforcement officials recorded. More often, monetary awards are presented by citizens detained on the spot of illegal logging or while transporting wood without proper documentation (i.e., the forest declaration, certificate of right to place, lease agreement or forest plantations purchase and sale contract, and monetary compensation).

The head of the forest police department, A. V. Sokolnikov, related the following in an interview posted to the internet:

> The system of bribes, in fact, is very well developed. All businessmen involved in the forestry business have worked together for a long time and know each other well. Money as a reward for a particular service is not transmitted in the form of cash – everything is much trickier. The custom is to give money to a third party. An official who performs organizational and administrative functions associated with the forest knows the entrepreneur and negotiates with the forest user: for example, ten percent of the cubic meter are required to be given to a third party. It is difficult for a law enforcement officer (especially in the police department) to prove it even technically, and a court case can unravel due to lack of evidence. Alternatively, transactions can be made through the account of the controlled firm. Because listening devices and recording equipment are used, money and gifts will never be discussed orally. If someone hints that he will "thank the firm for the service," his request will be denied on the spot. Entrepreneurs and bureaucrats know where to send the money, and how much. In short, they understand the price of the issue and of the business.[18]

In our region, the situation concerning bribe-taking and giving related to timber sales is a little easier. For example, on 19 February 2009, the Petrovsk-Zabaikalskii city court in Trans-Baikal sentenced an individual, "N," to one year probation. The crime was committed under the following circumstances: On 12 February 2009, "N," not wanting to be involved and found guilty for committing illegal logging, tried to pass banknotes of 5,000 rubles to "Z," knowing that he is the chief of public security police in the internal affairs department at the Petrovsky Zavod station. Moreover, "Z" is an internal affairs official who has the power to terminate the examination of "N"'s case (illegal felling of trees in the Maletinsk forestry precinct on 30.11.2009). However, the attempted bribe was intercepted by police officials who entered the official's room. Thus, "N"'s attempt to bribe an official was unsuccessful.[19]

According to an anonymous survey of employees of the State Forest Service, rewards are often offered for concealment of illegal logging, illegal transportation of timber, processing of permits, concealment of felling of extra amounts of wood, and for timber harvesting by citizens for their own needs with subsequent sales of timber in an amount far greater than indicated in the purchase and sale contract.[20]

Among cases that have received a great deal of public interest is the criminal case against officials of the Anti-Economic Crimes Departments of the Internal Affairs Directorate of the Region who as an organized group extorted bribes totaling more than 13 million rubles for 2008–2009 to approve exporters' applications (documents required by customs to transport timber across the border) and for forged titles establishing timber ownership. A corruption crime is usually committed for money, as confirmed by 96 percent of officials of State Forest Service and other law enforcement agencies surveyed.

Another aspect of the expansion of corruption in the forestry sector concerns remote areas of the territory. Officials of the State Forest Service are responsible for the protection of forest resources, but the territory they serve is vast and covered with taiga forests, especially in the northern districts of the territory. There are simply not enough employees to fully monitor and protect the forests. Foresters and staff in rural areas live in the same villages as the loggers. As a result, the forests are pillaged and valuable wood is smuggled abroad by gangs comprised of whole villages, where law enforcement and forest protection officials are simply unable to resist the bribes and/or threats of criminals. A commonly known concept is "to put on a bribe" (посадить на взятку). This term is used in criminal operations and it means when an official or a bureaucrat is given a bribe, he performs certain actions for the reward. In the future, he is offered more opportunities to continue this criminal activity and the official or bureaucrat cannot refuse or else he will be threatened with criminal liability. In this case, the corrupt official becomes a hostage of his own greediness; will he go to court for the bribes or continue to carry out the demands of the criminals? Law enforcement officials and forest rangers conduct raids, investigate places of illegal logging, and prosecute offenders. However, in circumstances of corruption, such raids have lost their preventive value, because they "catch" those with the illegal timber who do not collaborate with authorities or entrepreneurs who are trying to operate their businesses by legal means.

Thus, the core areas of forestry affected by corruption are: the issuance of forest declarations, the examination of cutting areas, the conclusion of lease agreements and purchase contracts for forest stands, conducting inspections and raids to detect illegal logging, transporting illegally cut timber, and issuing phytosanitary certificates.

We now turn to the issue of preventing and combating corruption crimes in the area of timber harvesting and trade. Figure 4.2 presents institutional, criminal-legal, and other measures that, according to a survey of law enforcement and forestry officials, are most effective:

- tightening the criminal law, which establishes liability for crimes of corruption;
- drafting a detailed legal regulation of relations arising during the sale of forest plantations, and use of forest areas;

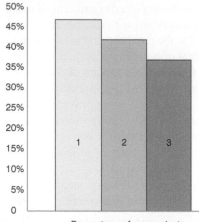

1. Stronger criminal legislation regarding the responsibility for corruption crimes

2. Detailed normative regulations in connection with the purchase or sale of forest sectors and/or the use thereof

3. Increased fines for corruption crimes that greatly exceed the illegal profits gained by the offender

Percentage of respondents

Figure 4.2 Most effective measures for combating corruption in the timber trade: opinions of law enforcement and forest officials in the Trans-Baikal Territory

- increasing the penalties for corruption crimes, many times greater than the benefits from the crimes.

The conditions that foster corruption crime include: extensive money circulation in wood resources' trade, high profitability of forestry business, the mentality of Russian citizens, low level of legal consciousness, the consumer attitude toward the environment, the lack of the federal and regional legislation, the broad authority of the officials of state and municipal governments, and the weak financial and social security of law enforcement and regulatory authorities' employees. During the interview with prosecutors of Trans-Baikal, we identified that one of the enabling conditions for corruption is lack of coordination between different control services: tax, customs, prosecutors and others.

To determine the causes of corruption crimes, it is necessary to identify the perpetrator and the key beneficiary of the crime. To protect forest resources from illegal looting, the interests of "black lumberjacks," forestry workers, and law enforcement agencies coincide. Why this is happening? Forest resources are owned by the state, but the Russian mentality is to consider the forests as common property; ownerless and and unnecessary to preserve, protect and restore.

Unfortunately, many individuals, including those responsible for protecting natural resources, strive to extract maximum profit from his/her official position. This results in a high level of bribery and other corruption-related crimes related to the distribution of forest resources. Employees of the Trans-Baikal State Forest Service and its territorial divisions have broad powers in the distribution of timber and forest land. Law enforcement officials can arrest the persons detained at the scene of illegal logging, and during the transporting or unloading of illegally harvested timber, and bring the perpetrators to justice.

Recommendations for combating corruption in the forest sector

There is a perception that private property is protected much better than state property because the state does not have the resources to ensure the safety of its property, specifically natural resources. If we convert forest lands from state to private ownership, there may be hope for the preservation of natural forest resources. However, there is also a risk that the owners of the forest will destroy the forest stands on their site in a short time period. It is therefore necessary to establish a price for transferring ownership of such forest sites, which could provide a return on forest resources for a long period of time. It is unlikely that someone will guard cheap natural resources. However, such measures are outlined rather loosely, and the development of a specific project requires the participation of many types of specialists, not only legal specialists.

Another innovative recommendation is to prohibit the sale of forest resources to individuals by giving this sector of the economy to the state. This would facilitate development of the national timber industry, but would contradict freedom of entrepreneurship. What is more important: the conservation of forests for our children or the padding of an entrepreneur's wallet? The answer to this question is obvious. As we pointed out, it is necessary to focus also on developing the legal awareness of forestry workers and of law enforcement and on promoting a widespread intolerance for all manifestations of corruption, as a phenomenon worthy of the attention of state and municipal employees, especially of law enforcement officers.

The following areas of activity are most prone to corruption: issuing forest declarations, examining of cutting areas, signing of lease agreements and purchase contracts for forest stands, conducting inspections and raids to detect illegal logging and transporting illegally cut timber. There are very few cases of extortion recorded. However, this is not an objective indicator because, according to law enforcement officials, the latency of corruption crimes reaches 45–50 percent and possibly even higher.

Undoubtedly, maximum transparency of logging companies and law enforcement officials and State Forest Service Agency officials will play a positive role. It is necessary to control not only for incomes, but for the expenditures of state and municipal officials and law enforcement officials. We hope that the ongoing work on combating corruption and organized crime will yield positive results.

Notes

1   I. A. Liasheva, "Illegal logging poses dangers to society," *Ekologicheskoe Pravo* 6, 2008, pp. 8–12.
2   M. Gritsiuk, "The woodcutters in the shadow," *Rossiiskaia Gazeta, Federal'nyi vypusk* no. 5144 (65) 30 March 2010.
3   According to materials of the document on the results of combating forest crimes in 2010 prepared by the Procuracy of the Trans-Baikal Territory. These materials are not published.

4   Ibid.

5   Ibid.

6   Presidential Order from April 13, 2010, N 460 "O Natsional'noi Strategii Protivodeistviia Korruptsii i Natsional'nom Plane Protivodeistviia Korruptsii na 2010-2011 gody," *Rossiiskaia Gazeta,* no. 79, April 15, 2010.

7   "Po Materialam Spravki o Rezultatakh Bor'by s Prestupleniiami v Sfere Lesopromyshlennogo Kompleksa za 2010 God," prepared by the Procuracy of the Trans-Baikal Territory. These materials are not published.

8   Zakon Zabaikalskogo Kraia, "Ob Organizatsii Deiatel'nosti Punktov Priema i Otgruzki Drevesiny na Territorii Zabaikalskogo Kraia," from 24 June 2009.

9   Zakon Zabaikalskogo Kraia "On Administrative Violations," 1 July 2009 No. 198-33K.

10 According to materials from the Procuracy of the Trans-Baikal Territory; especially a review of court and investigative practice for the year 2009–2010 contained in quarterly reports.

11 These crimes include: smuggling (188); Failing to return historical and archeological artifacts of the Russian Federation and foreign countries(190); and Trade in precious metals, natural precious stones, and pearls, (191). Ugolovnyi kodeks RSFSR, http://base.consultant.ru/cons/cgi/online.cgi?req=doc;base=LAW;n=126811;fld=134;d st=4294967295 (accessed 24 April 2012).

12 A. Petrov, "Lesnoi Kodeks Rossiiskoi Federatsii kak ob"ekt Korruptsionnosti," *Ustoichivoe Lesopol'zovanie,* 3(19) 2008, p. 3.

13 According to data from the survey of law enforcement officials and statistical materials provided by the Information Center of the Directorate of Internal Affairs of the Trans-Baikal Territory.

14 "O Protivodeistvii Korruptsii: Federal'nyi Zakon RF", 25 December 2008, No. 273-FZ, *Rossiiskaia Gazeta,* 30 December 2008, no. 266.

15 "Ob Antikorruptsionnoi Ekspertize Pravovykh Aktov i ikh Proektov: Zakon Zabaikal'skogo Kraia," 24 December 2008, *Zabaikal'ski Rabochii,* 12 January 2009, no.3.

16 "O Tselovoi Programme Zabaikal'skogo Kraia, 'Razvitie Lesopromyshlennogo Kompleksa Zabaikal'skogo Kraia (2009-2013)': Zakon Zabaikal'skogo Kraia," 26 September 2008.

17 "Zabaikal'skii Lesopromyshlennik Rasskazal o Korruptsii na Urovne Lesnichestv" http://www.wood.ru (accessed 5 February 2011).

18 Skazhem Korruptsii "Nyet!: Lesnaia Vziatka," (Elektronnyi resurs): http://www.for-expert.ru, (accessed 13 January 2011).

19 According to data from the official site of the Petrovsk-Zavodskogo suda Zabaikalskogo kraia http://www.petrovsk.cht.sudrf.ru (accessed 13 January 2011).

20 Data provided by the Procuracy of the Trans-Baikal Territory.

5 Environmental crimes in the territory adjacent to the petroleum-storage facility in the town of Kama in the Kambarsk Region of the Udmurt Republic

Larisa Pervushina

We must think about strengthening the responsibility for environmental crimes. However, the most sensible response is to develop more realistic methods of compensating damages and to require offenders to liquidate the pollution, including the most complicated forms of pollution, such as petroleum.

We need a plan of concrete activities and an organized team to prepare a package of normative documents. Finally, we need special registers and methods for monitoring regulations and providing effective solutions to various problems.[1]

D.A. Medvedev

Article 42 of the Constitution of the Russian Federation guarantees every person the right to a clean environment.[2] In accordance with legislation of the Russian Federation, [3] in protecting the population and the environment from the negative impact of waste products resulting from oil and oil production (during relocation, design, construction, reconstruction involving exploitation and exploiting objects of oil and gas extractive industries, objects of conversion, transportation, preservation and drilling of oil and gas and their byproducts), enterprises should clean and decontaminate the production waste. Oil, gas, and mineralized water must be decontaminated and dirty soil re-cultivated to alleviate the damage done to the environment during the exploitation phases.

Petroleum-related enterprises are prone to a high rate of accidents at every stage in the technical cycle, from extraction to preparation of consumable petroleum products. The majority of instances of environmental damage related to the petroleum industry result from non-standard technical business practices, and the bursting of main oil lines are considered to be extreme cases.

The surrounding environment is affected by massive petroleum spills. Depending on the quantity and type of oil or petroleum by-products spilled, abatement efforts may take anywhere from several months to several years – and occasionally decades, and require tremendous expenditures to restore the area to its original condition.

In the case of the petroleum-storage facility in question, its pollution of the surrounding environment has gone on for decades. Research completed in 1995

Figure 5.1 Evidence of environmental impact (photos by the author)

established that soil and groundwater had undergone lasting degradation, including contamination of the groundwater by dangerous petroleum-laced substances. The area subjected to contamination has continued to widen. The groundwater, soil, subsoil and air have experienced exceedingly high levels of contamination. There has been a shift in the types of vegetation that grow in the contaminated territory, and in several areas vegetation is simply dying off. Given the contamination levels, one can infer that seepage from the petroleum-storage facility has been constant for more than 15 years, and that the petroleum business which uses this facility has not undertaken the measures necessary to prevent seepage of petroleum by-products into the surrounding environment. (See Figure 5.1.)

The author suggests that the environmental contamination of the area adjacent to the petroleum-storage facility must be considered an environmental crime committed by the negligent petroleum business under Articles 250, 254, 261 of Chapter 26 of the Criminal Code of the Russian Federation.

Research methodology

Theoretical foundations

Oil and petroleum by-products are ubiquitous environmental polluters. In entering the environment, petroleum hydrocarbons begin to disrupt the ecosystem's balance over a very short term. The issue of preventing and mitigating petroleum contamination of soil and water is therefore crucial. Proper environmental management is particularly critical in areas harboring petroleum-business sites, where the production and disposal of dangerous substances not only creates an anthropogenic imprint on the area's ecosystem, but also poses a true public health threat.

Several academic works[4] detail the findings of research into soil contamination in various types of places: transportation-related areas (garages, filling stations near major thoroughfares, parking lots behind apartment buildings) and, as a control, the least contaminated soils in nature preserves. The most radical changes were observed in soils contaminated on the topsoil layer, which contains humus.

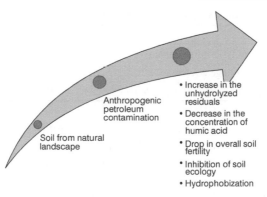

Figure 5.2 Anthropogenic soil contamination

Figure 5.2 reflects the fundamental understanding that all soils that have been subjected to anthropogenic contamination of oil and petroleum by-products reflect demonstrable decreases in humic acid – the foundation of soil fertility. Additionally, there was a dramatic increase in the level of unhydrolyzed residuals, or the portion of organic substances that is non-recoverable after the process of humus fractionation by various chemical extractants (which in the soils of natural landscapes is represented by humin and humin-like substances).

In climatic-soil areas and provinces the concentration of petroleum by-products in soil increases from south to north, from sandy soils to clay-like soils, from semi-moist to waterlogged, from treated to virgin.[5] In order to preserve and monitor the soil quality, as well as develop and choose technologies and methods for obtaining the product that would meet requirements, an assessment of the depth and degree of pollution is necessary.

The criteria for determining the concentration of oil and petroleum by-products in soils have already been established, which allows for easier soil management and abatement, increases public safety for those living in contaminated areas and restoration, and ensures the restoration of land to crop-output quality. The categorization of polluted areas[6] is undertaken in accordance with the *Procedural Guide for Determining the Damage from Chemical Contaminants* in Table 5.4, ratified by Roskomzem[7] on 10 November 1993 and by the Russian Ministry of the Environment on 18 November 1993. Contamination levels beneath 1000 mg/kg are considered permissible, while areas with levels of 1000–2000 mg/kg are considered mildly contaminated, those with 2000–3000 mg/kg are moderately contaminated, and those with more than 5000 mg/kg are severely contaminated. In accordance with the guide, the cost of damage abatement is borne by the enterprises, establishments, organizations and other legal entities responsible, and punitive costs are assessed based on the severity of contamination. Punitive damages are assessed by the Russian Ministry of the Environment and Roskomzem in accordance with the "Regulations on Conserving Degraded Agricultural Resources and Soils Polluted by Toxic Industrial Waste and Radioactive Substances," supported by Russian Government ordinance 555 of 5 August 1992.

Table 5.1 Chemical contamination of marshland by oil

Indicator	Decline (times)
Quantity of different types of groundcover	1.5–3.0
General number of foliage covers	6 and higher
Productivity of plant biomass of groundcover	10–36

In the instance of several entities bearing responsibility for pollution, punitive costs are assessed according to the particular level of responsibility of each.

As the territory of the subject research is marshland, the effects of oil and petroleum by-products on marshland were researched. (Table 5.1).[8]

Chemical contamination of marshes by petroleum and mineralized waters, in addition to the saturation of the territory, results in the alteration of the fundamental characteristics of the soil cover of marshland ecosystems. The quantity of types of groundcovers declines from 1.5–3.0 times, the general number of foliage covers declines by six or more times, and the productivity of the groundcover's plant biomass is 10–36 times lower than that of unspoiled marshland groundcovers. For example, the marshland cranberry crop on the surface oligotrophic marshes of the Middle Ob region is 56.1 berries per square metre. Oil extraction decreases the crop density per bog, as well as the bog area itself, resulting in a significant loss – from 38 percent to 100 percent – of cranberry output in marshland.[8]

Petroleum also affects the soil composition, although several types of microorganisms present in the soil might be cleansing organisms. Among the cleansing components might be petroleum-oxygenating microorganisms, aerobic gram-negative bacteria, and methane-oxygenating microorganisms.[9]

The contamination process damages soil irreversibly through deep changes in the soil composition resulting from the deterioration of its physiochemical properties and the absorption of petroleum by the soil. To better understand how soil is affected by petroleum it is first necessary to evaluate the role of the various components of petroleum in the overall petro-contamination process.[10] Light oil may have the following impact: in low concentrations it will not impact soil microorganisms, whereas in high concentrations it will negatively affect not only soil microorganisms, but also tall plants and microscopic soil organisms. In very high concentrations it will act as the fundamental substrate layer for hydrocarbon-oxidizing microorganisms.

Therefore, under circumstances where petroleum has seeped into soil, changes are expected in both the organic and inorganic soil compounds. These changes might result in reactions between the soil components and the petroleum or its destructive by-products, which might lead to negative changes in the natural soil composition.

The degradation of petroleum under the oxidizing conditions of surface-layer geosystems is a multi-stage, dynamic process made unique by the speed of permutations of the various components of the petroleum mix. The transformation of petroleum is dependent on the physiochemical and biochemical destructive and synthetic processes of transforming the hydrocarbon geo-substrata into a different heteroatomic substance with extraordinary geochemical potential.[11]

This research into petroleum's effect on environmental ecosystems is centered on identifying the principles of the ecosystems' transformation under the influence of exogenous factors, and is the theoretical foundation for studying petroleum contamination and effective restoration of ecosystems damaged by this contamination.

The research will reference analysis of water and soil samples which were taken as part of this study. The samples were taken by Samples Engineer Y. V. Nikitin on 19 November 2010 in accordance with Russian Government Standard 51592-2000, *General Requirements for Taking Water Samples*. The water samples were analyzed for petroleum-product content in accordance with *The Principles of Measuring the Mass Concentration of Petroleum Products in Environmental and Effluent Waters Using an Infrared Spectrometer* (Russian Federal Environmental Normative Document 14.1:2.5-95). Measurement of mass concentration of petroleum products was ascertained by extracting emulsified and dissolved petroleum by-products from water using carbon tetrachloride.

The soil samples were taken at a depth of 20 centimetres in accordance with the Russian Government Standards 17.4.3.01-83, *Environmental Protection: General Requirements for Taking Soil Samples,* and 17.4.4.02-84, *Environmental Protection: Procedures for Taking and Preparing Soil Samples.* They were analyzed for petroleum-product content in accordance with *The Principles of Measuring the Mass Concentration of Petroleum Products in Soils and Sediments Using an Infrared Spectrometer* (Russian Federal Environmental Normative Document 16.1:2.22-98).

The principle for measuring the concentration of petroleum by-products centers on the absorption of infrared output by petroleum hydrocarbon molecules. The concentration of petroleum by-products in the extracted sample is measured by calculating the relationship between the intensity of light flows with the measured and reference wavelengths traveling through the sample being studied.

The author also used a wide range of legal standards, published scholarly research, and local court practice in undertaking the subject research. The empirical basis of the research includes environmental crime cases, including those suspended and dropped; analysis of refuting documents; and interviews with officials in environmental agencies and those accused of or found guilty of environmental crimes.

Research process and findings

The objects of research were areas contaminated by petroleum and petroleum by-products and waterways in their vicinity, which were visited on foot by the author with the goal of visually ascertaining the boundaries of contamination.

The terrain chosen for study is flat, with almost no valleys or hills, with a very shallow slope toward the river Kama. The distance between the petroleum-storage facility and the Kama River is approximately one kilometre. The hazard to using waste levels on the given area as an example is that the terrain allows for active surface discharge into adjacent waters.

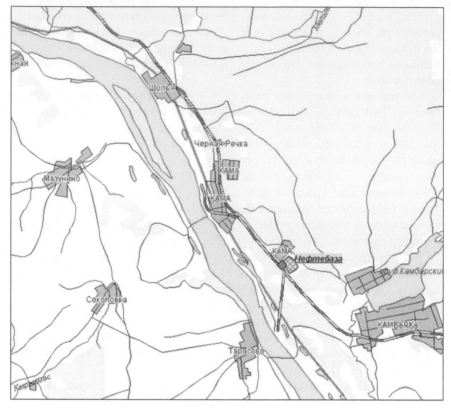

Figure 5.3 General map of the area studied

Figure 5.3 shows the Kambarsk petroleum-storage facility and the surrounding territory, which was the subject of the research.

The research examines the condition of environmental ecosystems located in the "sphere of influence" of the Kambarsk petroleum-storage facility, as well as waste-disposal areas (reservoirs, slurry pits, and residual dumping pools).

Analysis of technological stages

This section sets out the analysis of the main technological stages during storage and shipment of petroleum and petroleum products, and the identification of sources of environmental pollution at every stage of technological process.

The petroleum-storage facility is a complex for receiving, storing and off-loading petroleum and petroleum products. The selection of a petroleum-storage facility site must meet standards and requirements set forth in the following legislation:

- Sanitary rules and regulation # 2.2.1 / 2.1.1.1200 – 03 *Planning, Construction, Reconstruction and Exploitation of an Enterprise, Planning and Building in Populated Areas: Health and Safety Zones and Health Classification of Enterprises, Buildings and Other Such Objects;*

- Norms of technical planning of petroleum-related enterprises, as set forth by departmental standards of technological design 5-95. Volgograd, 1995 Russian Ministry of Fuel and Energy;
- Russian Federation Ministry of Energy Decree № 232 (19 June 2003) on *Establishing the Regulations of Technical Exploitation of Petroleum-Storage Facilities,* registered in the Russian Federation Ministry of Justice on 20 June 2003, № 4785; and
- Russian Government Committee for Industrial and Mining Safety Supervision Decree № 33 of 20 May 2003, *Procedures for Industrial Safety of Petroleum-Storage Facilities, Safety regulations* 09-560-03.

A site must be selected on the basis of factors other than simple economic advantage, which is principally the site's ability to keep transportation costs for petroleum and petroleum products as low as possible. The site chosen for a petroleum-storage facility must also conform to particular engineering requirements, particularly related to its geological and hydrogeological positioning.

The selected site must also preserve a required distance between its facilities and neighboring structures. One of the most important site-selection criteria is its proximity to major transport thoroughfares. The facility site itself, or an area in very close proximity, must have access to water and energy for housekeeping (cleaning/sanitation), production and fire-protection needs. The selected site must have a run-off slope for storm water as well as sewage water that will not present a threat to the neighboring communities. All petroleum-storage facilities are organized into zones for production-management as well as fire-safety reasons.

The zone dedicated to petroleum transport operations along the rail system contains structures for loading and unloading large shipments of oil and petroleum products. Other structures and amenities in this zone include:

- railroad sidetrack;
- loading-offloading trestles and platforms;
- various types of industrial pipeline;
- pumping stations to convey oil and petroleum products; and
- a control room for service personnel.

The zone dedicated to water transport of oil and petroleum products contains structures for loading and offloading large quantities shipped via waterways, including:

- a river cargo-handling pier;
- pumping stations;
- shoreline reservoirs;
- industrial pipelines; and
- a control room.

The storage zone contains:

- reservoirs;
- industrial pipelines;
- pumping stations; and
- a control room.

The work zone, in which petroleum products are batched into small shipments for loading onto oil tankers, containers and barrels, includes:

- trestles for loading oil into tanker trucks;
- areas for filling oil into barrels;
- warehousing for packaged petroleum products;
- a quality-control laboratory;
- packaging storage;
- a unit for packaging petroleum products for consumer sales; and
- a unit for recycling used oils.

The zone of structures necessary for servicing the facility includes:

- an engineering workshop;
- a boiler-house;
- a steam station;
- an electricity station or transformers;
- a unit for the production and maintenance of oil packaging equipment;
- water and sewer pipes;
- materials warehousing;
- a fuel storage unit for the facility operations; and
- fire-safety equipment.

Administrative offices zone, which includes:

- administrative offices for the petroleum-storage facility;
- a fire-safety depot;
- security personnel offices; and
- garages.

The scrubbing zone includes:

- an oil trap to separate the petroleum products from water;
- a retention pond for collecting industrial run-off; and
- a pumping station for the oil trap.

The largest threat to the environment from the petroleum-storage facility comes from several elements of the warehousing and loading facilities. Oil and petroleum products are highly flammable, have a low ignition threshold, are able to conduct electricity, create pyrophoric substances from sulfur, and are able to self-ignite on contact with air. Hydrocarbon gases are explosive-prone and noxious, 3–4 times heavier than air, can collect and remain in low-lying areas (ditches, wells, pits, ravines) for an extended period.

The industrial activities of the petroleum-storage facility affect all parts of the biosphere: the atmosphere, the hydrosphere and the lithosphere.

To identify the fundamental ways in which the petroleum-storage facility affects the environment, we must examine the activities of the facility in each of the zones described above.

Industrial waste, which contains petroleum by-products, has the most damaging effect from the petroleum storage and shipment processes on the natural environment. Each stage in the industrial process produces its unique type of waste containing petroleum by-products and become pollutants of the adjacent territory.

Geo-ecological analysis of the territory

Environmental assessment

Environmental assessment of the soil and plant covers includes overland and underground waters, and air quality. To identify the contamination's focal points and establish their boundaries, the author conducted an assessment of the entire territory encompassing the petroleum-storage facility's zone of environmental influence.

The walkthrough observation included an overall environmental quality assessment and a landscape indicator study, during which the author identified anthropogenic changes to the landscape in question. Additionally, key tracts of land where industrial processes had the most environmental impact were surveyed. The routes taken in conducting this assessment are mapped in Figure 5.4.

Groundwater samples were taken and sent for quantitative analysis to the accredited physical-chemistry laboratory of the Administration of the Ministry of the Environment of the Udmurt Republic.[12] Each sample sent for chemical analysis was 1.5 litres, comprised of three half-litre glass bottles with rubber corking, as required.

Thirty-five soil samples were taken from the area contaminated by petroleum products and also sent to this accredited laboratory.

The points where soil and groundwater samples were taken are depicted in Figure 5.5. The soil samples are marked with black circles and the water samples with white circles. Samples were taken in areas that appeared to display signs of the most severe contamination, as well as in areas that did not appear to have been contaminated at all.

Soils in the area of analysis were sandy-podzolic soils, boggy-podzolic soils and boggy soils, and sandy by mechanical composition. Sandy soils are characterized by low buffering capacity and high water permeability. Contaminating substances appearing in a soil cover with a sandy particle composition damage the area's fundamental ecology, easily penetrating and collecting in the soil, groundwater, and nearby waterways.

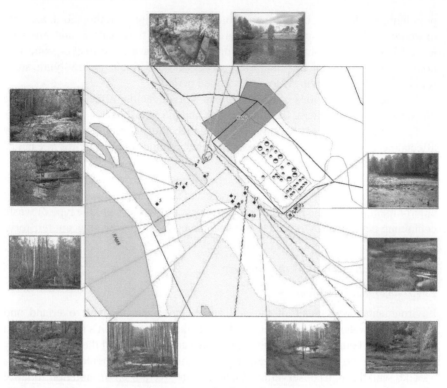

Figure 5.4 Map of the routes taken in conducting research

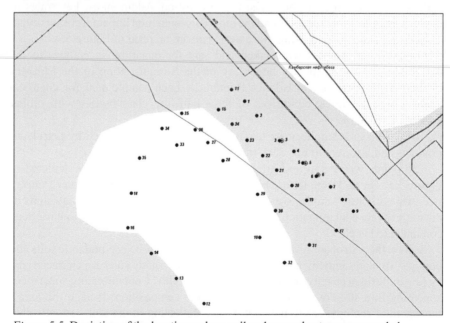

Figure 5.5 Depiction of the locations where soil and groundwater were sampled

Figure 5.6 Satellite image of the area researched

Soil contamination generally occurs through petroleum spillage on its surface, forming a more or less continually spreading stain of petroleum products with heightening concentrations into lower-lying areas. Additionally, even terrain, a high level of groundwater, and delayed clean-up facilitate a wide spread of petroleum products and penetration of the contaminants into the deep soil profile (based on visual field inspection research).

In conducting the subject research, the author analyzed synthesized color imagery of the Kambarsk petroleum-storage facility taken by the QuickBird satellite from a height of 72 meters from the area (taken in answer to a query for imagery of the territory). For illustration, images from an internet geodata portal are provided in this chapter. Spectral analysis was performed using the specialized computer software ENVI 4.5.

The image of the territory adjacent to the petroleum-storage facility was taken on 14 July 2007. The satellite image (Figure 5.6) was utilized to identify large areas of polluted forest cover stemming from anthropogenic impact of the storage of oil and petroleum products at the Kambarsk facility, as well as to ascertain the boundaries of the affected forest areas. Three main "contamination zones" were identified, as highlighted in Figure 5.6.

The darkest stain, the "epicenter," of zone 1 depicts a small contaminated area: it aligns with the flow of petroleum from the facility, and represents an area of intense anthropogenic impact on the environment.

Table 5.2 Analysis of groundwater samples from contaminated areas

№	Sample location	Results of analysis		
		Maximum residue limit mg/dm³	Petroleum products mg/dm³	Average (in times)
1	Drill-hole №6	0.05	1.27	25.4
2	Drill-hole №5	0.05	6.4	128
3	Drill-hole №3	0.05	4.95	99

The zones shown in Figure 5.6 are in the form of concentric areas emanating from the darkest areas, or epicenters, of contamination. The epicenters are depicted by blackened circles, while the areas of mid-level contamination surround these epicenters.

The routes taken in conducting the field research and sampling the soil and water were informed by satellite imagery to ensure that the author investigated the most polluted areas.

Chemical analysis of water and soil samples

Water and soil samples taken from the contaminated areas adjacent to the petroleum-storage facility underwent quantitative chemical analysis to determine their petroleum-product content. Groundwater samples were analyzed according to fishing standards of a maximum residue limit of $0.05mg/dm^3$ and approximate safety levels of substance impact for water and watershed areas used by the fishing industry.

The results of the groundwater analysis are displayed in Table 5.2 .

The water samples substantiate that the quality of groundwater does not satisfy the requirements of established norms.

Petroleum contamination levels of water and soil were evaluated by calculating the level at which the samples exceeded the baseline concentration of petroleum products (50 mg/kg – sample point 26).

Significant contamination of the territory by petroleum products is demonstrated by the results of the analysis, as detailed in Table 5.3.

Evaluation of the condition of the contaminated area

The level of contamination of the subject area was evaluated on the basis of the field research conducted, bolstered by the chemical analysis of the water and soil samples, as well as past research on the subject area (available through the Government Foundation for Geological Information of the Republic of Udmurtia).

Soils and groundwater are considered contaminated when their concentration of petroleum products reaches a level at which negative environmental changes occur: the natural balance of the soil ecosystem is disrupted, the microorganisms in the soil begin to die, plants begin to lose their crop productivity and/or die, the soil morphology or hydrophysical soil features change, soil fertility drops, underwater and groundwater sources are endangered by the breakdown of petroleum products from the soil or water or their absorption into water.

Table 5.3 Analysis of soil samples from contaminated areas

Sample point	Results of analysis		
	Maximum residue limit mg/dm^3	Petroleum products mg/dm^3	Average (in times)
1	50	80000	1600
2	50	2040	40.8
3	50	27400	548
4	50	900	18
5	50	14700	294
6	50	96	1.92
7	50	164	3.28
8	50	140	2.8
9	50	98	1.96
10	50	880	17.6
11	50	80	1.6
12	50	74	1.48
13	50	116	2.32
14	50	117	2.34
15	50	65	1.3
16	50	96	1.92
17	50	74	1.48
18	50	62	1.24
19	50	1810	36.2
20	50	21700	434
21	50	800	16
22	50	82	1.64
23	50	137	2.74
24	50	193	3.86
25	50	510	10.2
26	50	Less than 50	-
27	50	2830	56.6
28	50	1800	36
29	50	137	2.74
30	50	176	3.52
31	50	259	5.18
32	50	101	2.02
33	50	62	1.24
34	50	176	3.52
35	50	283	5.66

Determining the level of soil contamination is vital to deciding whether to take soil-rehabilitation measures. The threat level for soil contamination is considered that at which soil can no longer rejuvenate itself.

In consideration of the physical-geographical conditions of the subject area (the climate, soil types and content, vegetation, etc.), as well as the exogenous land-use factors that affect the soil's ability to rejuvenate itself, the following scale was used to determine at which points to take soil-rehabilitation measures:

- under 50mg/kg: uncontaminated
- 1000–2000 mg/kg: mildly contaminated
- 2000–3000 mg/kg: moderately contaminated
- 3000–5000 mg/kg: severely contaminated
- above 5000 mg/kg: very severely contaminated.

Mildly contaminated soil areas can be rehabilitated within two to three years, whereas moderately contaminated areas can take four to five years. Serious environmental damage to soil begins at the level of petroleum concentration above 3000mg/kg: at this concentration petroleum products begin to seep into underground water and considerably disrupt the ecosystem balance. Research shows that when petroleum contamination extends to an area of more than 30.8 hectares, the area must be recultivated.

The field research undertaken indicates that the petroleum contamination of the subject territory occurred over time from the flow of petroleum products out of the Kambarsk petroleum-storage facility and into adjacent ground water, and along railway beds where it comes to the surface and spreads along a wide area.

Following from these observations, it will be necessary to reconstruct the Kambarsk petroleum-storage facility's reservoirs in order to fully eliminate the contamination source. Without rebuilding the reservoirs, measures undertaken to recultivate the subject area will be ineffective.

Calculating environmental damage

The following section focuses on calculating the environmental damage done by dumping liquid petroleum waste into the terrain around the industrial complex in defiance of environmental regulations resulting in economic and environmental damage to the area.

Contemporary scholarship on environmental damage in Russia and abroad has formulated a theoretical and methodological basis – and has crafted a set of instruments – for quantifying the cost of environmental damage.[13] These instruments are used quite broadly in international practice to quantify environmental damage and seek recompense, including through court proceedings.

The main problems encountered in estimating costs and assessing fines for environmental damage include:

- the utilization of estimation standards that employ incorrect indicators of environmental damage;

- the failure of these measures to value the socio-economic consequences of environmental damage (public health consequences and private property damage);
- the lack of methodological tools to calculate the costs of various types of environmental damage, such as water, mineral deposit cavities (tapped mines), city soils, the ecosystems of nature preserves, vegetation not included in forest land, rare and endangered animals and plants, as well as sub-types of environmental damage (such as subtypes of damage resulting from petroleum contamination).

Environmental damage is defined as the negative changes to the surrounding environment, triggered by degradation of natural ecosystems and the depletion of natural resources, resulting from its contamination. However, the terms "degradation of natural ecosystems" and "depletion of natural resources" are not defined by law, which leads to various interpretations of the term "environmental damage" and hampers efforts to determine its extent. This is most complicated by the contentious issue of *precisely when* degradation of natural ecosystems occurs.

Moreover, this definition does not encompass all possible consequences of the subject contaminant and, in particular, societal consequences. Documents outlining standards lack a standard definition of the terms "harm," "damage," and "losses." Economic literature and academic textbooks use terminology such as "losses averted," "direct losses," and "collateral damages," and use different methods for calculating each. In the Civil Code of the Russian Federation, the term "damage" is broadest and encompasses real losses (involving loss of profit or income) as well as moral damages. The term "losses" is typically applied more broadly to mean more than simply material losses and approaches the meaning of the term "damage."

The Russian Federation Law "On Protecting the Environment" contains the most general principals of valuing and seeking compensation for environmental damage sustained due to violations of environmental laws and regulations. It is sensible to allow the law "On Protecting the Environment" and the Civil Code of the Russian Federation to govern the calculations of environmental losses, inasmuch as these documents set forth general principles of valuation and compensation as described above.

According to Article 15 of the Civil Code of the Russian Federation, losses are considered to include those costs associated with restoration of the damaged property, recompensing monetary losses, or restoring the rights that were violated ("real losses"), as well as unrealized income due to the infringement ("denied profit"). Real losses are measured as the cost of the lost property, while denied profit is calculated by unrealized income that the proprietor would have received under the normal course of business without the subject infringement. This article thereby defines and cements as legal norm a fundamental economic formula used fairly broadly to calculate material and other related costs associated with crimes against property and resources, including those occurring in nature.

The application of this formula to valuation of *environmental damage* signifies that expenses associated with erasing all traces of the violation must include expenses for restoring the natural environment to its original quality, and must be consistent with the principles of calculating 1) real damages, and 2) denied profits. Real damages relate to the price of destroyed or damaged natural objects or environmental components, and denied profits are all unrealized income from the area damaged or destroyed.

The principle discussed is utilized in assessing damage to the environment as a whole, land or forest resources, wildlife and its sanctuaries, nature preserves, water bodies, etc. The Russian Federation Law "On Protecting the Environment" defines the responsibilities related to full restoration of damaged or destroyed areas, as well as the procedure for determining compensation amounts for environmental damage resulting in violations of environmental-protection laws and regulations.

In reviewing these laws and legal regulations, the following measures must be taken to eliminate the negative impact on the environment from the Kambarsk Petroleum-Storage Facility:

1   removal and recycling of all accumulated petroleum by-products and waste;
2   recultivation of destroyed or damaged areas with the goal of minimizing environmental degradation;
3   reconstruction of the storage facility's reservoirs.

Analysis of international methods

International best practices in monitoring environmental degradation

To retain objective information concerning the creation, implementation and evaluation of environmental policies, there must be an adequate system of environmental monitoring. The distribution of environmental monitoring duties among the national government, states (regions) and local authorities depends on the institutional structures in each country. Local-level authorities generally exercise control over this arena: they should monitor the emissions from factories and take action when limits are exceeded or other infractions occur, such as when a factory is lacking the required permits for its activities. If permits are not as yet required for a specific type of industry, then local authorities must ascertain whether the factory's activities meet standards and regulations in its sphere of activity.

Large industrial accidents leading to environmental crises provided the impetus for passage by the European community of acts to set environmental-safety standards.

There is a burgeoning interest in putting the onus on the enterprise to submit proof to the regulating authorities that they have identified all existing environmental dangers, have taken all necessary environmental precautions, and have informed all their own personnel and nearby communities of their industry and its potential environmental impact. There is a growing pan-European database of industrial polluters, which includes incidents related to the polluters, the state of the territories in which they work, the cause of accidents, and their experience in prevention and elimination of contamination. Goals, standards and regulation procedures have been established that govern the state of the environment, monitoring of its condition, and concrete restrictions.

Federal assistance to disadvantaged territories, industries and enterprises is becoming more common in a variety of countries. Germany adopted a number of measures to improve the lowlands along the Rhine and retrofit chemical enterprises along its banks. A reduction in pollution was achieved through repurposing and closing industrial polluters, retraining and finding alternative employment for displaced workers, and relocation of inhabitants.

A positive example of utilizing federal resources to improve the environment occurred in the Great Lakes region of Canada and the United States.

One of the most difficult aspects of overcoming environmental crises remains the assignment of responsibility, first for criminal conduct leading to a crisis situation, and second for environmental crimes leading to emergent conditions. In a market economy, material responsibility is a vital element.

Material responsibility requires a specific enterprise to answer for specific contamination that has inflicted specific damage – not for contamination of an entire territory or according to the principle of "the polluter must pay." Citizens who find themselves in terrible environmental situations often bring class-action suits against polluters. The number of such lawsuits is not particularly high and the rate of prosecution is low. However, those that do succeed might bankrupt the industry they sue and the very existence of class-action cases might increase the pollution-prevention measures that various industries take.

Several countries' legislation requires the creation of environmental compensation funds by companies whose industry may be hazardous to the environment or may pose an environmental threat. The funds are spent on environmental rehabilitation efforts in the event that the guilty parties become unable to afford them, or if the concrete polluting agent cannot be established.

Insurance claims can be similar to these environmental compensation funds, although insurance pay-outs from environmental damage are uncommon and this type of insurance coverage is not widely accessible.

With some trepidation, governments are beginning to give incentives for environmentally-sound industrial practices. In these arrangements, a government will grant tax deductions for certain types of environmental activities: the use of zero-emissions technologies, the use of clean technology, the donation of valuable natural habitats to government agencies or civil-society organizations, and the forfeiting of property rights to areas that are national parklands or nature preserves. However, there is a growing concern that these arrangements instill in industry an over-reliance on government hand-outs to "do the right thing": good environmental stewardship ought to be a fully-integrated business concept for any industry and its responsibility to the citizenship of the country in which it operates, irrespective of any subsidies or privileges offered.

The European Union's environmental-protection laws are the most complete and well-established. These laws are to a significant degree "direct-response legislation" and therefore, in contrast to the framework laws of the Russian

Federation, they do not require further interpretation using acts handed down by other government agencies. Moreover, the EU system of laws covers nearly all fundamental issues of protecting and utilizing the natural environment. And, most importantly, the EU legal norms resulted from an inclusive dialogue among civil society, government and business.

The governing principle of the EU environmental-protection legislation, set forth in Directive 96/61/EC and in its most recent version, 2008/1/EC of 15 January 2008 entitled "Integrated Pollution Prevention and Control Directive," requires a constant reduction in the degree of environmental impact. With the goal of creating a balance between the requirement to minimize pollution and practical technical capabilities, the Directive envisions the application of a mechanism to calculate indicators of environmental impact on the principle of "best available techniques (BAT)." The term "BAT" acknowledges that the subject technique is the best from the standpoint of meeting environmental standards and is accessible to entities seeking to apply it. The Directive notes that the requirement to utilize BATs applies only to the largest sectors of the economy and to industries that have the most significant impact on the environment.

The goal of the Directive is to provide an integrated approach to protecting the environment through a comprehensive system of administration and monitoring over industrial and manufacturing processes. The key element of the approach is a general principle set forth in Article 3 of the Directive: operators (manufacturers) must take all necessary measures to prevent polluting the environment, in particular through the adoption of BATs, which will allow them to increase their environmental efficiency.

The functioning Russian environmental-protection laws fail to combine the use of monetary sanctions for excessive environmental pollution with the introduction of targeted parameters of acceptable environmental impact for manufacturers. Nor do these laws include some sort of significant incentives for instituting environmentally-sound measures into manufacturing or using green technologies.

One of the biggest priorities for environmental policy in the medium term, in addition to efforts to mitigate environmental crises, should be to adopt the principle of applying BATs to neutralize manufacturing's environmental footprint. This transition to practices exemplified by developed countries, where neutralizing the environmental footprint is achieved by establishing technical standards for users of environmental resources that are linked to the BATs, would be in the best interest of improving Russian manufacturing's market competitiveness.[14]

Minimizing negative environmental impacts of specific technologies is important at both the project level, where parameters are enforced, and at the level of determining the parameters of acceptable environmental impact. The latter is a key issue: the government agencies responsible for oversight and standard-setting must understand that technical specialists of the manufacturing industry are stakeholders, and they must include these specialists when establishing standards and promoting policies to reduce negative environmental impact from manufacturing. The EU's experience in dramatically reducing pollution of its environment clearly exemplifies this notion.

It is important to note that neither increasing the liability of legal entities nor increasing the fines for environmental contamination have been effective without the added dimension of using economic triggers to make environmentally-sound business practices rational and in the best interest of industry.

Work must be done to improve environmental monitoring as a whole. It may be possible to create a public–private partnership to establish a system for industry to track the health of the environment and thereby receive certain benefits, such as lowering of payments made to offset environmental damages or other such incentives.

Investigation and prosecution of environmental crimes

Among the various types of environmental crime, industry-related violations of environmental-protection laws are particularly heinous. The Criminal Code of the Russian Federation, enacted on 1 January 1997, is the first criminal legislation in the country to contain an entire chapter entitled "Ecological Crimes." Included among environmental crimes are activities that influence the environmental order and environmental security of society as a whole. However, criminal law standards that establish responsibility for environmental crimes are rather complicated, as they require significant knowledge and experience concerning the crime, including that gained through forensic investigations. The paltry number and state of criminal cases in the environmental sphere are testimony of the low level of effectiveness of efforts to prevent, identify, and investigate these types of crime. One disturbing element to this is the high latency and low level of identification of environmental crimes. Only 3.1 percent of all environmental crimes are identified and registered as such. [15] Due to ineffectual investigations, in many cases the causes of the environmental crimes are not determined, the responsible parties are not identified, and the necessary compensations for damages are not made. It is not rare that sentences for guilty verdicts are lowered for unfounded reasons in cases where individuals have committed dangerous environmental crimes. [16]

Many scholars, including A. I. Vinberg, V. E. Konovalova, E. I. Maiorova, N. T. Malakhovskaia, G. A. Matusovskii, A. P. Rezvan, E. R. Rossiskaia, M. V. Saltevskii, A. R. Shliakhov, and others, have written on the particular issues of investigating criminal cases opened as a result of infringements of environmental laws. The scholars mentioned above have researched the use of various types of specialized knowledge, the necessity of which arose during the investigation of a number of environmental crimes that affected lifestyles within communities. These cases yielded a number of recommendations to increase the success of investigations into environmental crimes, as follows:

- create a singular methodology for investigating environmental crime cases, including specialized detection and investigation techniques;
- improve existing methods of taking samples for court cases to ensure quality-control and the adherence of sampling to best practices;

- finish the creation of both a team of specialists prepared to testify in environmental-crimes cases, and the scholarly-methodological base documents for court cases; and
- rectify the lack of clarity in the provision of environmental appraisals for the court system; this haziness is due in part to the fact that specialists' opinions vary as to which principles of environmental safety are most fundamental for court cases.

According to data compiled by the Administration for Combating Violations of Environmental Laws in the Republic of Udmurtia (hereafter referred to as the "Environmental Administration of Udmurtia"), in its first five years of existence more than 500 environmental crimes and more than 18,000 infractions were identified. Total damage to nature from these identified violations was valued at 75 million rubles. In the first three months of 2010 alone the Environmental Administration of Udmurtia identified 59 crimes, of which 24 were illegal hunting and illegal logging or removal of shrubbery. The majority of crimes discovered are connected to non-payment of pollution offsets.

Analysis of the adherence to requirements

Requirements are outlined in "The Principles of Industrial Safety of Petroleum-Storage Facilities and Other Warehouses of Petroleum Products, and the Assignation of Responsibility and Punishments." Violations related to the petroleum industry occur annually in the Republic of Udmurtia; for example, on 25 March 2010, a break occurred on an unsanctioned 18-kilometre pipeline in the republic's Alnash region. The resultant catastrophic spill covered a 4.2 hectare area of farmland; information about the spill was kept from the public and the producer-proprietor never disclosed the volume of the spill. The Russian Environmental Monitoring Administration's chapter in the Republic of Udmurtia carried out an unplanned inspection of the facility involved in this incident, following which they implemented environmental clean-up activities and levied fines against the guilty parties.

It must be noted that the majority of environmental crimes discovered in the petroleum sphere have been accidents, and of a catastrophic level. However, small leaks of oil and petroleum products (relative to the size of spills from pipeline accidents) can continue unnoticed for a long time. We might consider the situation in the subject region surrounding the Kambarsk petroleum-storage facility to be this type of instance, wherein the continuous contamination results in degradation of the area's environment.

One of the factors delaying the management of this problem in Kambarsk is the lack of standard-setting documents and directives that establish concrete criteria for evaluating the level of soil and groundwater pollution from oil and petroleum products, as well as the absence of economic and environmental justifications for adopting certain methods of reversing the damage to soil cause by catastrophic oil incidents, as informed by foreign and domestic experience.

Currently the media – television, radio, periodicals and the internet – have the most active and pressing influence on society and politicians. The illumination of environmental problems and environmental crimes has a positive effect – it galvanizes environmental movements, government monitoring agencies and society at large. Environmental oversight agencies, in their turn, begin to more aggressively inspect unscrupulous users of environmental resources, levy fines, and determine administrative and criminal punishments.

Environmental policy is the policy of the future. Environmental policy demands one thing: that all agents of the government, including municipal-level officials, respect and abide by the governing laws of the country and the Constitution of the Russian Federation. Article 42 of the Constitution establishes the right of every citizen to a wholesome environment. Individuals who manage territories or areas must not forget about ecology. Unfortunately, often such individuals in today's reality think only about business and money and forget about the environment.

Notes

1 D.A. Medvedev, at the meeting of the Presidium of the State Council on Issues of Improving Environmental Regulations in Russia, 27 May 2010.
2 The Constitution of the Russian Federation, http://www.constitution.ru/index.htm (accessed 9 April 2012).
3 "On Protecting the Environment," Federal Law (FZ) no. 7, 10 January 2002.
4 See, for example, V. I. Larionov, "Modeling of oil flow during the decompression of oil pipelines," *Voprosy bezopasnosti ob'etkov neftegazovogo kompleksa: Sbornik nauchnykh trudov,* Moscow: TsIEKS, 2004,14–21; L G. Bakina, E. E. Orlova, and N. E. Orlova, "Change in the humic condition of urban soil under the influence of oil pollution," Congress report, "Innovation in the Ecology and Security of Life Activity," St. Petersburg, vol. 2, no. 3, 2000, 185–187, and M. V. Begak "NDT: Effective, accessible, productive," *EKO-biulleten' InEkA* 2009, no. 3,134.
5 S. A. Bogoliubov, "Extraordinary environmental situations," *Ekologicheskoe pravo*; G. P. Lapina et al., "Physical-Chemical Characteristics of Environmental Pollution from Technogenic Catastrophes (Oil Spills)," *Khimiia i khimicheskaia bezopasnost* 1, 2007, 24–32; M. A. Prokasheva, "Protection and rehabilitation of soil contaminated by oil and petroleum products," *Agrokhimicheskii vestnik* 2, 2000, 27–29.
6 Bogoliubov, "Extraordinary environmental situations."
7 Translator's note: this official government agency translates as Russian Land Committee.
8 Lapina et al., "Physical-Chemical Characteristics."
9 V. S. Feoktistova, "Hydrocarbon activity of soil organisms during the process of biodegration of highly viscous oil," *Problemy regional'noi ekologii* 8, 2000, 185–186.
10 Prokasheva, "Protection and Rehabilitation of Soil."
11 Ibid.
12 The accreditation number for the laboratory, valid through 27 July 2011 (and valid at the time of testing in 2010) is POOC RU 0001.511868.
13 V. D. Ermakov and E. N. Zhevlakov, "The status of environmental crime in the Russian Federation," *Kriminal'nologicheskie i pravovye problemy obespecheniia ekologicheskoi bezopasnosti. Sbornik nauchnykh trudov,* Moscow, 1996, 7–17; "Methods for Calculating Damages as a result of State Violations of Water Laws," PD33-5.3.01-83; and "The order for determining the scale of the damage from chemical contamination," stipulated by Roskomzemom 10 November 1993 and

Minpriroda RF 18 November 1993; and GOST (State Standard) 17.5.1.02-85, "The Classification of Damaged Soil for Recultivation."

14   Begak "NDT: Effective, accessible, productive."

15   V. I. Smirnov, P. I. Meshkov, and E. M. Zadorozhnaia, *Geologicheskoe stroenie i gidrogeologicheskie usloviia mezhdurech'ia Izha i Kamy*, vol. I and II, Dzerzhinsk, 1973, 1–39.

16   Sanitary rules and regulations 2.1.4.1074-01, on drinking water.

6 The impact of metallurgical and cement industrial waste on Central Russia's environment

Elena Bocharnikova

Heavy metals (HM) are considered as the most hazardous environmental pollutants because they are prone to cause serious harm in human organisms. Heavy metals enter the human body through the intake of food and water as well as by inhaling dust. Many HM accumulate in living organisms over time and pose serious health problems. The most dangerous metals to human health are mercury, cadmium, lead, arsenic, copper, tin, and zinc.[1]

The iron and steel, cement and chemical industries are characterized by large volumes of annually produced waste containing HM, radionuclides, and other contaminants. Available information about the levels of certain contaminants in industrial waste is either incomplete or deliberately understated by the metallurgical and cement-producing industries. In order to obtain certificates authorizing the use of silicon-containing waste in agriculture and construction, contaminants must be shown to be low. However, the metal and cement industries take advantage of the shortcomings of Russian legislation with respect to industrial waste disposal. Rigorous standards to control the ecological dangers posed by HM waste do not exist.

In Russia, the main law governing the need for compliance with environmental and human health regulations is the Constitution of the Russian Federation (1993).[2] Article 42 guarantees the right of Russian citizens to a healthy environment and to reliable information about it. In addition to the Constitution are the laws on "Environmental Protection" (1992)[3] and the Federal Law "On Ecological Expertise" (1995),[4] Metallurgical and cement industry waste is regulated by Federal Law "On the Production and Consumption Waste" (1998).[5]

Currently, the definition of "waste utilization" is found only in the "GOST (Government Standard) 30772-2001: Resource-saving. Waste Management. Terms and Definitions" (waste utilization: the activities associated with the use of waste at the stages of the technology cycle, and /or the secondary use or recycling of discarded products, which is applied in part, but does not contradict federal law). As shown in Table 6.1, the GOST introduces a four-point based waste risk assessment system in Russia. However, as practice shows, the classification scheme in Table 6.1 neither accounts for waste pollution nor obligates enterprises to process waste in a way that will ensure environmental safety. Unfortunately, a legal framework restricting the waste utilization or requiring companies to detoxify metals and cement containing HM waste in practice does not exist.

Table 6.1 Classification of danger from chemicals substances in DL$_{50}$*

Value K$_{\Sigma}$, determined according DL$_{50}$	Class of toxicity	Degree of danger
Less than 1.3	I	Extremely dangerous
From 1.3 to 3.3	II	Highly dangerous
From 3.4 to 10	III	Somewhat dangerous
More than 10	IV	Minimally dangerous

*DL$_{50}$ (dosis letalis 50) is a dose of an agent (substance, bacteria, viruses, toxins and others) that has a lethal effect on 50% laboratory animals.

Metallurgical and cement industries have always produced large amounts of waste because of the technologies they use. The production process entails the extraction of metal from ore. Key stages of this process are ore processing, metal smelting, and then re-melting certain conditions. The first two processes yield the most metallurgical waste. It should be noted that ore contains a large amount of silicon compound.[6] The enrichment and smelting processes involve the separation and extraction of the metal- and silicon-containing substances. The percentage of silicon content in the ore is often higher than the content of the desired metal. As a result, gigantic waste piles accumulate and large dumps of waste are found at most large metallurgical plants.

As early as 1881 in the United Sates it was suggested for the first time that industrial waste could be used as a silicon fertilizer.[7] Utilization of large amounts of silicon-containing industrial waste was possible by integrating it into the soil to improve fertility and increase plant resistance to adverse weather conditions.[8] Another application of the waste was to use it as gravel for road construction.

The history of silicon fertilizers and the use of silicon-containing industrial waste as soil ameliorants can be divided into several time periods. From the early twentieth century to the mid-1950s many studies of silicon concluded that it protected plants from fungal diseases. The experiments were conducted most aggressively in Japan, where empirical data showed the effectiveness of silicon fertilizer and soil ameliorants. As a result, in 1955 a governmental decree requiring the use of silicon fertilizer for growing rice was published in Japan.[9] Metallurgical waste products also have been used in Japan.

The period from the mid-1950s to the late 1990s was characterized by theoretical research, which established the impact of active forms of silica on the soil–plant system. At the same time, in some countries (USA, Brazil, Australia) large-scale experiments were conducted on the use of silicon-containing wastes in agriculture.

Since 2000, a sharp increase has been seen in the use of silicon fertilizers. This can be explained by the accumulation of knowledge regarding the positive effect of silicon on the soil–plant system, as well as by a new appreciation for traditional fertilizers and crop-protecting chemicals. The use of silicon-containing compounds in plant growing allows the reduction of the doses of conventional fertilizers and pesticides and at the same time increase the yield of crops.

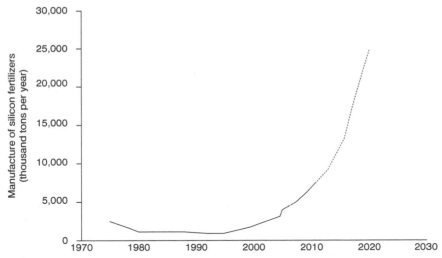

Figure 6.1 Growth in use of industrial waste with silicon for agricultural purposes (1975 to 2010) and the forecast for 2020

Therefore, silicon fertilizers found in industrial waste have been used much more intensively in recent years. As a result, today more than 7 million tons of metallurgical waste is recycled world-wide. (Figure 6.1). Continued growth of the use of silicon-containing compounds in agriculture is expected, in particular with regard to the effects of climate change. The optimal silicon plant nutrition increases the resistance of agricultural crops to moisture deficit and excess salt in the soil.[10]

Research methodology

One of the purposes of our research was to evaluate the content of HM in waste as one of the key components posing risks to the environment and human health. Initially, we undertook a study of metal and cement industry waste to evaluate the possibility of using some of this waste as fertilizers and soil silicon ameliorants.[11] Three ways for obtaining the samples of industrial by-products were used: a) samples obtained from factory waste on site, b) samples sent to the author from other factories, and special exhibitions and c) samples of metallurgical and cement industries' waste collected in dozens of plants in various regions of Russia over the course of three decades, from 1984 to 2010.

An examination of the waste samples identified the total amount of HM in accordance with standard methodology,[12] the X-ray fluorescence method to analyze solid samples,[13] and the atomic adsorption method with an alkaline sintering and subsequent dissolution of solid samples.[14] Analysis of waste by the X-ray fluorescence method was performed in the geology department of Moscow State University during the years 1986–1990. The analysis of waste using the atomic-adsorption method was performed at the Institute of Physical, Chemical and Biological Problems of Soil Sciences, Russian Academy of Sciences, between 1991 and 2010. All analyses were performed in three-stage replication.

The environmental risk level of the waste was determined by comparing the results of the total share of HM in metal and cement industrial waste samples to the maximum levels of permissible concentrations (PDKs) of chemical substances in waste and environmental spheres.[15] A list of metal and cement companies was taken from the website http://www.zawod.ru/.[16]

Information about the modernization of the enterprises was obtained from the companies' websites or during site visits. To assess possible chemical contamination of the environment by metallurgical and cement waste, we used the following criteria:

1   presence and level of HM as compared to the maximum permissible concentration (MPC);[17]
2   availability of equipment for waste storage (tents, special storage facilities); and
3   availability of special equipment for processing (sorting, packaging).

The following levels of possible chemical contamination were identified:

1   low
2   average
3   high
4   very high.

To evaluate the risk of chemical pollution by metallurgical and cement industries waste geographically we analyzed the distribution of plants according to the federal districts of the Russian Federation.[18] Figures were drawn using the data presented in Tables 6.3–6.5.

Table 6.2 Key criteria for evaluating possible chemical contamination of the environment by metallurgical and cement industrial waste

Grade	Level of concentration	Equipment available for waste storage	Equipment available for processing waste
Low	Nonoccurrence of heavy metals	Special storage reservoirs	Special unit for waste processing
Average	HM content 2–5 times lower than PDK	Canopies	Waste packaging
High	HM content approaching maximum permissible concentrations	Equipped open space	Use of sprinkling system for prevention of dust accumulation
Very high	HM content higher than maximum permissible concentrations	Lack of equipment	Lack of equipment

Waste contamination and the law

The Ministry of Natural Resources and Ecology of the Russian Federation posted a draft of the Federal Law "Amendments on the Federal Law on Industrial and Consumer Waste" on 23 August 2010.[19] Although the draft law improves the procedure of transferring waste from one company to another, it fails to address the need to protect citizens' rights to a healthy environment and to provide them with reliable information about the state of the environment. This enables industrial enterprises to violate the Russian Constitution with impunity. One example of unauthorized soil contamination is Novolipetsk Metallurgical Complex, one of the four largest steel companies in Central Russia. More than 10 million tons of waste are stocked on the site and nearly 60 million cubic metres of wastewater are dumped into the River Voronezh annually.[20] The chemical composition of soil around the metallurgical plant in Volgograd contains extremely dangerous levels of lead (2975 mg/kg) and arsenic (1002 mg/kg). In addition, elevated concentrations of mercury (0.1 mg/kg), zinc (520 mg/kg) and cobalt (55 mg/kg) were detected.[21]

In the Belgorod region, the largest polluters are the joint stock companies "Oskolcement," "Belgorod cement," "Shebekinsk Chalk Plant," and "Building Stroymaterialy." These enterprises account for 21 percent of the emissions of all pollutants in the region. The emissions of construction companies alter the geochemical environment in large areas adjacent to the factories. There are farms in the contaminated areas where agricultural crops are grown.[22]

The study of the impact of the cement industry on the environment of Volsk city (Saratov region) reveals that the concentration of HM (in excess of the background content) by hazard class are as follows:

- first hazard class:
  - cadmium – 3800
  - lead – 6.92
  - zinc – 3 39
  - arsenic – 9.53
- second hazard class
  - cobalt – 108.8
  - copper – 6.5
  - antimony – 91.63
  - chromium – 3.79
  - molybdenum – 34.1
- third hazard class
  - strontium – 11.3.[23]

It is clear that there are risks of chemical pollution at other metallurgical and cement industries as well, which will inevitably harm human health. That is why this evaluation of chemical hazards is very important.

Many waste by-products contain a large amount of HM. For instance, the waste of Cherepovets Metallurgical Plant in Volgograd oblast was used as a road cover, although it contains nickel, chromium, lead and cadmium.[24] Recycling techniques of HM waste should include a detailed risk assessment of soil pollution. Waste containing large amounts of HM and other pollutants should be purified before they are applied or stored in special landfills. However, the environmental protection standards developed in the Soviet Union are no longer enforced. Environmental policy in the USSR was part of the general social economic policy of the Communist Party. These laws were mainly for setting government standards (GOST) for content of pollutants in various wastes.[25] Today it has become common practice to ignore compliance with the established rules and regulations.

Most of the silicon-containing industrial waste in Russia is stored next to the enterprises. Millions of tons of silicon-containing waste are often stored without any hermetic sealants, such as isolation hydrofuge (a waterproof adhesive for thermal insulation XPS panels). Therefore, mobile forms of HM together with precipitation penetrate the groundwater and become a source of contamination of groundwater and surface water. For example, we encountered environmentally toxic waste storage practices while visiting the Voskresensk Chemical Plant (Moscow Region) in 1986. Several million tons of silicon-containing wastes with a high content of chromium, copper, zinc, and tin are stored in the factory site.

Major metallurgical plants produce large quantities of fine dust consisting of oxides of different elements. The latter is trapped by gas treatment equipment and then either fed into the sludge collector or sent to further processing. These wastes are characterized by a large proportion of iron and are used by the same metallurgical plant in most cases.

Metallurgical waste

Data about the risk of environmental pollution by individual plants of metallurgical industry in Russia is presented in Tables 6.3 and 6.4. The results revealed that more than 40 percent of Russian metallurgical plants have a very high risk of chemical contamination of the environment (Figure 6.2). At the same time, only 17 percent of plants have a low risk of chemical contamination of the environment. Most of the metallurgical plants of Russia are located in the Urals Federal District (36 percent), followed by the Volga Federal District (20 percent) and the Central Federal District (17 percent) (Figure 6.3).

Our analysis of the distribution of metallurgical plants with a very high risk of chemical contamination of the environment reveals that the Urals Federal District (38 percent), where most of the metallurgical plants with outdated equipment is concentrated, is in first place, followed by the Northwestern Federal District (23 percent) and the Siberian Federal District (15 percent) (Figure 6.4). The risk of contamination is lower in the Central and Volga Federal Districts, where most metallurgical plants have upgraded their technologies.

The waste of metallurgical plants contains a large amount of HM (Table 6.3). This includes lead, cadmium, zinc, nickel, and copper. At plants such as the Lysvensk Metallurgical Plant (Urals Federal District) and Sulinsk Metallurgical Plant (Southern Federal District) the waste consists of uranium and other radionuclides. Because these wastes are radioactive, they require special attention.

Table 6.3 Russian metallurgical enterprises containing silicon-rich by-products

Enterprise	Major pollutants	Type of by-products	Region
Central Federal District			
Kosogorsk Metallurgical Enterprise	Ni, Hg	Steel slag	Tula region
Lipetsk Metallurgical Enterprise "Free Falcon"	Zn, Cu, Co	Steel and pig-iron slag, dust	Lipetsk region
Metallurgical Enterprise "Electrosteel"	Pb	Steel and pig-iron slag, dust	Moscow region
Oskolsk Enterprise Metallurgical Mechanical Engineering	Cu	Steel and pig-iron slag, dust	Belgorod region
Shchelkovsk Metallurgical Enterprise – Shchelment	Pb	Steel slag	Moscow region
Northwestern Federal District			
Lisvensk Metallurgical Enterprise	Radionuclides	Steel and pig-iron slag, dust	Perm region
Novgorod Metallurgical Enterprise	Hg, Pb, Co	Slag	Novgorod Region
Chusovskoi Metallurgical Enterprise	Ni, Cr, Cd	Steel and pig-iron slag, dust	Perm region
Southern Federal District			
Sulinsk Metallurgical Enterprise – STAKS	Zn, U	Steel and pig-iron slag, dust	Rostov region
Taganrog Metallurgical Enterprise (Tagmet)	Hg, Pb, As, F	Steel slag	Rostov region
Urals Federal District			
Ashinsk Metallurgical Enterprise	Cu, Ni, Zn	Steel slag	Cheliabinsk region
Verkhnii-Saldinsk Metallurgical Enterprise	As, F, Su	Steel slag	Sverdlovsk region
Verh Isetsk Metallurgical Enterprise "Emal-Manufacturing"	Zn, Cd, Pb	Steel and pig-iron slag, dust	Sverdlovsk region
Zlatoust Metallurgical Enterprise	Cu, Zn, Ni	Steel slag	Cheliabinsk region
Kamensk-Uralsk Metallurgical Enterprise	Cu, Ni, Zn	Steel and pig-iron slag, dust	Sverdlovsk region
Magnitogorsk Metallurgical Enterprise	Pb, Cd, Ni, Zn	Steel and pig-iron slag, dust	Sverdlovsk region
Nizhneserginsk Metallurgical Enterprise	Cd, Ni, Zn	Dust	Sverdlovsk region
Revdinsk Metallurgical Enterprise -	Cd, Zn, Ni	Slag and Dust	Sverdlovsk region
Serovsk Metallurgical Enterprise	Cu, Zn, Ni	Steel slag	Sverdlovsk region

Continued…

Table 6.3 continued

Enterprise	Major pollutants	Type of by-products	Region
Ufaleisk Enterprise Of Metallurgical Mechanical Engineering	Cu, Zn	Steel and pig-iron slag, dust	Cheliabinsk region
Cheliabinsk Metallurgical Enterprise (Mechel)	Pb, Cd, Ni, Zn	Steel and pig-iron slag, dust	Cheliabinsk region
Volga Federal District			
Volgograd Metallurgical Enterprise "Red October"	Cu, Ni	Steel slag	Volgograd region
Viksunsk Metallurgical Enterprise	Cd, Pb, Ni	Steel slag	Nizhnii Novgorod region
Gorkovsk Metallurgical Enterprise (GMZ)	Pb, Cd, Ni, Zn	Steel slag	Nizhnii Novgorod region
Kulebaksk Metallurgical Enterprise "Ruspolymer"	Cu, Ni	Dust	Nizhnii Novgorod region
Samara Metallurgical Enterprise	Pb, Zn, Ni	Dust	Samara region
Omutinsk Metallurgical Enterprise (OMMET)	Zn, Cu	Pig-iron slag, dust	Kirov region
Siberian Federal District			
Guriev Metallurgical Enterprise (GMZ)	Cu, Cd	Steel and pig-iron slag, dust	Kemerov region
Krasnojarsk Metallurgical Enterprise – Kramz	Pb, Cd, Zn	Steel and pig-iron slag, dust	Krasnojarsk region
Novosibirsk Metallurgical Enterprise Kuzmin	Cd, Ni, Cu	Dust	Novosibirsk region

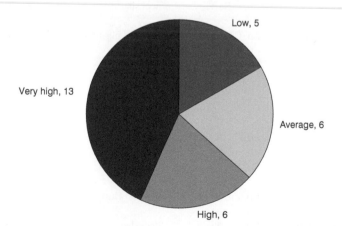

Figure 6.2 Distribution of Russian metallurgical plants by risk of chemical contamination

Table 6.4 Russian metallurgical enterprises: possible pollution levels

Enterprise	Type of waste utilization	Level of treatment and landfilling	Possible level of environmental pollution
Central Federal District			
Kosogorsk Metallurgical Enterprise	Storage, building material	Low	High
Lipetsk Metallurgical Enterprise "Free Falcon"	Storage, building material, soil amendment	Low	Very high
Metallurgical Enterprise "Electrosteel"	Storage, building material	Availability of special landfill	Average
Oskolsk Enterprise Metallurgical Mechanical Engineering	Storage, building material, soil amendment	Average	Average
Shchelkovsk Metallurgical Enterprise – Shchelment	Storage, building material, soil amendment	Average	Average
Northwestern Federal District			
Lisvensk Metallurgical Enterprise	Storage, building material, soil amendment	Availability of special landfill	Very high
Novgorod Metallurgical Enterprise	Storage and utilization	Average	Very high
Chusovsk Metallurgical Enterprise	Storage, building material	Availability of special landfill	Very high
Southern Federal District			
Sulinsky Metallurgical Enterprise – STAKS	Storage, building material, soil amendment	Availability of special landfill	Very high
Taganrog Metallurgical Enterprise (Tagmet)	Storage, Building Material, Soil Amendment	Availability of special landfill	Low
Urals Federal District			
Ashinsk Metallurgical Enterprise	Storage, building material	Low	High
Verkhnii-Saldinsk Metallurgical Enterprise	Storage, building material, soil amendment	Average	Average
Verkhsetsk-Metallurgical Enterprise "Emal-Manufacturing"	Storage, building material, soil amendment	Availability of special landfill	Very high
Zlatoust Metallurgical Enterprise	Storage, building material, soil amendment	Low	High
Kamensk-Uralsk Metallurgical Enterprise	Storage and utilization	Low	Very High

Continued…

Table 6.4 continued

Enterprise	Type of waste utilization	Level of treatment and landfilling	Possible level of environmental pollution
Magnitogorsk Metallurgical Enterprise	Storage, building material, soil amendment	Low	Very High
Nizhneserginsk Metallurgical Enterprise	Storage and utilization	Low	Low
Revdinsk Metallurgical Enterprise	Storage and utilization	Average	Very high
Serovsk Metallurgical Enterprise	Storage, building material, soil amendment	High	Low
Ufaleisk Enterprise Of Metallurgical Mechanical Engineering	Storage and utilization	Average	High
Cheliabinsk Metallurgical Enterprise (Mechel)	Storage, building material, soil amendment	Availability of the special landfill	Very high
Volga Federal District			
Volgograd Metallurgical Enterprise "Red October"	Storage, building material, soil amendment	Average	Average
Viksunsk Metallurgical Enterprise	Storage, building material, soil amendment	Average	High
Gorkovsk-Metallurgical Enterprise (GMZ)	Storage, building material	Availability of special landfill	High
Kulebaksk Metallurgical Enterprise "Ruspolymer"	Storage and utilization	Low	Average
Samara Metallurgical Enterprise	Storage and utilization	Low	Low
Omutinsk Metallurgical Enterprise (OMMET)	Storage, building material	Availability of special landfill	Very high
Siberian Federal District			
Guriev Metallurgical Enterprise (GMZ)	Storage, building material, soil amendment	Availability of special landfill	Very high
Krasnojarsk Metallurgical Enterprise – Kramz	Storage, building material, soil amendment	Availability of special landfill	Very high
Novosibirsk Metallurgical Enterprise Kuzmin	Storage and utilization	Average	Low

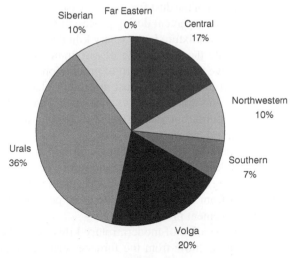

Figure 6.3 Distribution of metallurgical plants by federal district

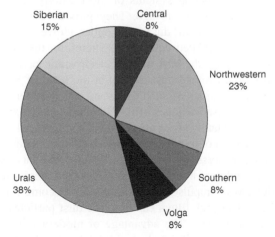

Figure 6.4 Distribution of metallurgical plants with a very high risk of chemical contamination

Cement waste

Back in the mid-1990s, cement waste, the annual volume of which is estimated at 20 million tons, was fully utilized as a soil ameliorant. However, this waste is classified as toxic in the United States and Western Europe, and is subject to storage in specially equipped dumps, because it may contain such heavy HM as cadmium, zinc, nickel, and lead.

Cement dust has a detrimental impact on the environment in a radius of several kilometres from the plant because it is the finest high-grade cement fraction. Results from a study of particulate dusts generated during the manufacture of Portland cement clinker, indicate that isolated from sources of dust pollution –

polydisperse.[26] The content of the dust fraction increases from less than 10 microns in the material to 10.75 to 75 percent during processing. Most fine dust is formed during the firing of the raw mixture in the rotary kiln drying production process.

Cement plants, despite the wide range of raw materials and manufacturing equipment, generally have the same pattern of production.

All the technological units emitting dust at cement plants are equipped with dust collection devices. This not only allows a significant amount of semi-finished or finished product to return, but also prevents dust pollution in the air around cement plants and adjacent territories.

Dust in the atmosphere from cement plants is formed mainly by three sources of dust release: rotary kilns, cement mills, and silos. The main source of dust emission is clinker baking kilns. In most cases, the amount of dust emitted into the atmosphere with gases from furnaces makes up to 80 percent of the total amount of dust released during cement production.

During the normal operation of modern rotary kilns using the wet method of clinker production, dust outflow from the furnace relative to the weight of dry material entering the furnace, is typically 5–8 percent.[27]

Currently, dust collection systems of most enterprises use electrical filters installed twenty or more years ago. They provide a purification rate of 95–98 percent or 300–800 mg/m^3 dust at the outlet. However, some enterprises lack such filters and use outdated equipment. Therefore, they must be modernized. To date, however, only the best foreign electrostatic precipitators (ESPs) with 5–7 fields provide a residual dust content at the level of 50–100 mg/m^3. At the same time, the dimensions of such filters are much larger than those currently used. The essential disadvantages of ESPs include complexity of the design, the vicissitudes of a constantly changing chemical and physical environment, and the residual electrification of trapped dust particles that does not allow them to return to the production process. As a technical system, ESP has reached the limits of its development and does not meet current, more strict emissions requirements.

Sleeve filters with impulse regeneration may become a good alternative for electrical filters. The sock filters can hold the dust particles that are larger than the holes in the materials. The advantage of modern fabric filters is based on several factors. While classifying the potential danger of cement plants as sources of pollution we considered the presence or absence of bag filters.

Waste from the cement industry has been used in agriculture for a long time and primarily as a source of potassium and calcium on acidic soils.[28] In addition, cement dust is an efficient silicon fertilizer. To calculate the possible contamination of soil and natural waters by HM from storage areas or places where recycled cement industry waste is used, please refer to the list of plants contained in Tables 6.4 and 6.5.

Analysis of the data showed that more than 50 percent of cement plants in Russia are characterized by a high risk of chemical contamination of the environment (Figure 6.5). At the same time only 10 percent of cement plants in Russia are characterized by a low probability of chemical contamination of the environment. Most of the cement factories in Russia are located in the Central

Table 6.5 Cement enterprises of Russia with silicon-rich wastes

Enterprise	Major pollutant elements	Type of by-products	Region
	Central Federal District		
Belgorod Cement Plant	Pb	Dust	Belgorod region
Voskresensk Cement, Cement Plant – Lafarge	Pb, Zn, Cd	Dust	Moscow region
Lipetsk Cement Plant	Zn	Dust	Lipetsk region
Mihailovsk Cement Plant	Cu, Zn	Dust	Riazan region
Podolsk Cement Plant	As, Pb	Dust	Moscow region
Shchurovsk Cement Plant	Hg, Cu, Pb	Dust	Moscow region
	Northwestern Federal District		
Vorkytinsk Cement Plant	Hg, Cd, Pb, As	Dust	Komi republic
Savinsk Cement Plant, Eurosenet Group	Zn, Cu	Dust	Arhangelsk region
Pikalevsk Cement Plant	Pb	Dust	Leningrad region
Cesla-Slansevsk Cement Plant	Zn, Cu	Dust	Leningrad region
	Southern Federal District		
Novorosijsk Cement Plant	Cd, Ni, Cr	Dust	Krasnodar region
	Urals Federal District		
Katav-Ivanovsk Cement Plant -	Cr, Ni	Dust	Cheliabinsk region
Magnitogorsk Cement Plant	Hg, As	Dust	Cheliabinsk region
Nevianovsk Cement Plant	Cr, Co, Pb	Dust	Sverdlovsk region
Sukholozsk Cement Plant	Cu, Ni	Dust	Sverdlovsk region
Uralcement, Korkinsk Cement Plant	Zn, Ni	Dust	Cheliabinsk region
Cheliabinsk Cement Plant "Master Kraft"	Zn, Pb, Cu	Dust	Cheliabinsk region
	Volga Federal District		
Volsk Cement Plant	Cd, Pb, As, Co	Dust	Saratov region
Gornozavodsk Cement Plant	As, Hg	Dust	Perm region
Zhigulevsk Cement Plant	Hg, Pb, Cu	Dust	Samara region
Novotroitsk Cement Plant	Zn, Ni	Dust	Orenburg region
Pashijsk Metallurgical-Cement Plant	As, Co, Cr	Slags, dust	Perm region
Serebriakovsk Cement Plant	Cu, Ni	Dust	Volgograd region
Ulianovsk Cement Plant	Hg, Pb, Cu	Dust	Ulianovsk region
	Siberian Federal District		
Angarsk Cement Plant (Siberian Cement)	Hg, As, Ni	Dust	Irkutsk region
Iskitimsk Cement Plant	Cu, Zn	Dust	Novosibirsk region
Kyznetsk Cement Plant	Ni, Pb, Cr	Dust	Kemerovo region
Topkinsk Cement Plant	Co, Zn	Dust	Kemerovo region
	Far Eastern Federal District		
Teploozersk Cement Plant	Hg, Pb, Cu	Dust	Jewish Autonomous Region

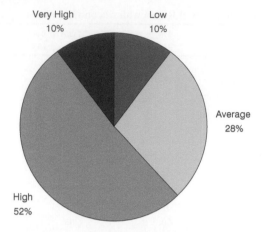

Figure 6.5 Ratio of cement factories in Russia in terms of possible chemical contamination of the environment

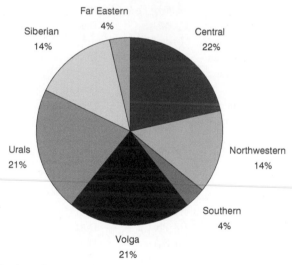

Figure 6.6 Distribution of cement plants by federal districts

Federal District (22 percent), followed by the Volga Federal District (21 percent) and the Urals Federal District (21 percent) (Figure 6.6).

Thirty-three percent of the cement plants in Federal Districts operate with outdated equipment, followed by analysis of the distribution of cement plants with very high and high Federal District (33 percent), with a high concentration of cement plants with outdated equipment, is in first place followed by the Central Federal District (17 percent), the Urals Federal District (17 percent) and the Siberian Federal District (17 percent) (Figure 6.7).

Cement plants' waste is less polluted by HM compared with the waste of metallurgy enterprises. However, they also may contain lead, cadmium, zinc, nickel, and copper.

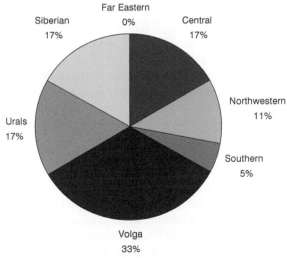

Figure 6.7 Distribution of cement plants with high and very high levels of possible chemical contamination of the environment by federal district

Conclusions

Analysis of the data shows that the waste from metallurgical and cement industries can contaminate soil, water, and air. We have examined the available data on the content of HM in the waste of metallurgical and cement industries and have collected waste samples of metallurgical and cement industries. Samples of the silicon-containing wastes were analyzed for heavy metal content of the first-, second- and third-class danger according to Russian government standards (GOST) by using the method of atomic adsorption. As a result, data on the content of HM in industrial waste were systematized. We also identified potentially dangerous areas of chemical pollution due to violations of the law or imperfections in the Russian legislation on metallurgical and cement industries' waste storage and utilization.

It is clear that controls over the content of HM in the steel and cement industries' waste must be regulated and tightened. This can be achieved in part through mandatory certification of materials that will be used in road construction, agriculture and other sectors of the economy. Environmental control over the content of HM (especially over their mobile forms) in waste storage areas is mandatory.

Appendix 1: The impact of toxic chemicals on humans

Mercury

Mercury is a *thiol poison* that disturbs protein metabolism and the enzymatic activity of the body. It is toxic to humans of almost any physical condition and is known for its wide range of harmful effects. Along with the poisoning, mercury and its compounds affect the sex glands, impact embryos, and cause malformations and deformities and lead to genetic changes in humans. Some research suggests that inorganic mercury can cause cancer.[29] Mercury has especially strong impact on the nervous and excretory systems.

Organo-mercuric compounds lead to severe impairment of the central nervous system wherein nerve cells can be completely destroyed. The compounds can also cause muscular and limb pain, impaired vision and hearing, and speech disorders. These physical impairments are practically irreversible and require prolonged treatment.

In addition, the impairments appear after months or even years after human exposure to methyl mercury. The high toxicity of low doses of which passes through biological membranes penetrates the brain and spinal cord, the peripheral nerves, pass through the placental barrier and accumulates in the fetus. Mothers who have had mild mercury poisoning, give birth to children with cerebral palsy, because the prenatal period is the stage of the life cycle susceptible to this metal. In nursing mothers methyl mercury passes into breast milk and then into the blood of infants.[30]

Cadmium

Cadmium is one of the most toxic HM. Acute food poisoning by cadmium occurs when one receives a large single dose with food (15–30 mg) or with water (13–15 mg). Signs of acute gastroenteritis – vomiting, pain and cramps in the pit of the stomach – are evident. The inhalation of cadmium dust or vapor can result in pulmonary edema, headache, nausea or vomiting, algidity, weakness and diarrhea are symptoms of cadmium poisoning and death has even been recorded as a result of such poisoning.[31]

Lead

Lead enters the human organism from food and water and dust particles. It is most dangerous is when it enters the body via dust from contaminated soil, as the increase in lead content in the soil for every 100 mg/kg causes an increase in toxicant concentration in the blood at 0.5–1.6 mg/dl. The toxic effects of the lead occur first in the central nervous system. The toxic effects of the lead occur first in the central nervous system. and are manifest in headaches, fatigue, memory impairment, and autonomic dysfunction. After lead enters the body, it penetrates the blood plasma and quickly connects to red blood cells, causing a disorder of the porphyrin metabolism, heme synthesis and activation of anaerobic glycolysis, increasing platelet aggregation. Inhibiting the enzyme heme synthesis, lead causes hypochromic anemia, a marker of which is protoporphyrin.[32]

Arsenic

Arsenic is known is known to have one of the highest pathology rates, because nineteen pathologies are associated with this element. Several hundred tons of arsenic would be sufficient to poison most of the human race. Arsenic's toxic effect on the human body varies depending on the dose and duration of exposure. Symptoms of acute intoxication include nausea, vomiting, stomach pain; symptoms of chronic intoxication are manifested in fatigue, muscle pain, prostration. Acute and chronic intoxication are accompanied by lethargy, headache, confusion, convulsions.[33]

Copper

Copper plays an important role in maintaining normal blood composition, however, an excess of copper compounds in the blood stream could have the opposite effect and cause diseases such as anemia and harm the functioning of the respiratory tract and liver.[34]

Tin

Tin may cause mental illnesses like psychosis, hysteria, emotion imbalance and other ailments.[35]

Zinc

Zinc compounds have an effect on the metabolism of copper and iron, causing them to breach. Excessive amounts of zinc in the body may (according to experimental data) have a carcinogenic impact as well as toxic effects on heart, blood, gonads, etc.[36]

Appendix II: Waste samples and classifications

Wastes of metallurgical enterprises and cement plants were obtained in the period from 1984 to 2010 by personal visits by V. V. Matychenkov to the factories of: Metallurgical Enterprise "Elektrostal" (Moscow region), Shchelkovsk Metallurgical Enterprise – Shchelmet (Moscow region), Kosogorsk Metallurgical Enterprise (Tula region), Voskresensk Cement Plant (Moscow region) in 1984 by V. V. Matychenkov; Taganrog Metallurgical Enterprise (Rostov region), Magnitogorsk Metallurgical Enterprise (Cheliabinsk region), Volgograd Metallurgical Enterprise (Volgograd region), Belgorod Cement Plant (Belgorod region), Podolsk Cement Plant (Moscow region), Chelyabinsk Cement Plant (Cheliabinsk region) in 1985; Ulianovsk Cement Plant (Ulianovsk region), Zhigulevsk Cement Plant (Saratov region) in 2007.

Samples sent to investigators for examination

Kuznetsk Cement Plant (Urals Federal District), Samara Metallurgical Plant (Volga Federal District), Kulebaksk Metallurgical Plant (Volga Federal District), Ashinsk Metallurgical Plant (Urals Federal District), Sulinsk Metallurgical Plant

(Southern Federal District), Chusovsk Metallurgical Plant (Urals Federal District), Lipetsk Metallurgical Plant (Central Federal District), Sebriakovsk Cement Plant (Volga Federal District), Volsk Cement Plant (Volga Federal District), Novorossiisk Cement Plant (Southern Federal District), Slavchevsk Cement Plant (Northwestern Federal District), Shchurovsk Cement Plant (Central Federal District) and Novgorod Metallurgical Plant (Northwestern Federal District).

Waste classifications

Industrial waste can be classified according to various features, including:

1   industry – ferrous and nonferrous metals, ore – and coal mining, oil and gas industries, etc;
2   phase composition – solids (dust, sludge, slag), liquids (solutions, emulsions, suspensions), gases (carbon oxides, nitrogen oxides, sulfur compounds, etc.)
3   production cycles – during the extraction (smelting) of metals from raw materials, during the enrichment (tailings, fine pulps, overspills), in pyrometallurgy (slag, sludge, dust, gases), in hydrometallurgy (solutions, sludges, gases).

Metallurgical plants with a closed cycle (cast iron, steel-mill products) solid wastes can be of two types – dust and slag. Slag is an industrial waste which is generated during high-temperature processes. As a silicon fertilizer, slag from ferrous and non-ferrous metallurgy and aluminum production, and slag from the phosphate industry are used.

The main product of blast-furnace smelting is cast iron. By-products are slag and blast furnace gas. On average, in the combustion of one ton of dry coke 3400 m^3 blast furnace gas with an average calorific value 3.96 MJ/m^3 is produced. Dust and gaseous emissions from blast furnaces are formed as a result of complex physical and chemical processes. It is believed that with blast-furnace gas contains dust coming with the charge (formed during the crushing burden materials, mainly coal), and dust which is formed by dragging melting stock column in a blast furnace.[37]

The mass of dust, introduced with furnace gases is 20–100 kg/t of cast iron. The average concentration of dust in blast gas is 9–55 g/m^3, but under machine damage the size can be increased up to 200 g/m^3. The amount of blast furnace gas produced is 3880 m^3/t of wet coke, or 4000 m^3/t of dry coke, or 2000–2500 m^3 per ton of cast iron. The composition of this dust may include HM that can be scattered around the blast furnace, resulting in chemical pollution of soil and water.

Contamination of the soil and water occurs most often during the utilization of slag. The production of one ton of cast iron yields 0.9 tons of slag. Annually ferrous metallurgy produces 71 million tons of slag in Russia alone.[38] It has been shown that from this amount 23 million tons, or 32 percent, can be used as a soil ameliorants.[39] The rest of the slag goes to dumps or to the production of construction materials, where they can become a source of contamination with HM.

Notes

1 K. Reilly, *Metal Pollution of Food*, Moscow: Agropromizdat, 1985.
2 Russian Federation Constitution, *Rossiiskaia Gazeta*, 25 December 1993.
3 Law on "Environmental Protection," *Bulletin of the Congress of National Deputies of the Russian Federation*, 10, 1992, 457.
4 Law "On Ecological Expertise" N174-FL, *Rossiiskaia Gazeta*, 23 November 1995.
5 Law "On the Production and Consumption Waste," *Rossiiskaia Gazeta*, 24 June 1998.
6 K. A. Cherepanov, G. I. Chernish, V. M. Dinelt, and J. I. Sykharev, "Utilization of Secondary Material Resources in Metallurgy," *Metallurgiia*, Moscow, 1994, 10, 224.
7 J. Zippicotte, J. Zippicotte Fertilizer, Pat. USA N238240, *Official Gazette of the United States Patent Office*, 19(9), 1881, 496.
8 V. V. Matichenkov and E. A. Bocharnikova. "Use of Metal Industry Waste for Improving Plant Phosphate Nutrition and Increasing Plant Drought Resistance," *Agrokhimiia* 5, 2003, 42–47.
9 F. Ma and E. Takahashi, *Soil, Fertilizer, and Plant Silicon Research in Japan*, Amsterdam: Elsevier, 2002.
10 Matichenkov and Bocharnikova, "Use of Some Metal Industry Wastes."
11 V. V. Matichenkov, "Amorphous Silicon Dioxide in Soddy-Podzolic Soil and its Influence on Plants," Dissertation Abstract, Moscow State University, 1990; Matichenkov and Bocharnikova "Use of Some Metal Industry Waste."
12 Maximum content of toxic substances in industrial waste for determination of the toxicity levels (UVT. AS USSR 27.12.1984, Ministry of Health USSR 18.12.1984 N 3170-84); J. Zippicotte, J. Zippicotte Fertilizer, Pat. USA N238240, *Official Gazette of the United States Patent Office*, 19(9), 1881, 496.
13 A. A. Kommisarov and S. B. Andreev, "Roentgen-Fluorescent Method of Analysis." *Method Learning For Laboratory Work*, St. Petersburg State University, St. Petersburg, 2008.
14 Russian Federation Constitution, *Rossiiskaia Gazeta*, 25 December 1993.
15 Maximum content of toxic substances in industrial wastes for determining toxicity levels (UVT. AS USSR 27.12.1984, Ministry of Health, USSR 18.12.1984 N 3170-84).
16 *List of Russian Federation enterprises*, http://www.zawod.ru/ (accessed 29 November 2011).
17 Maximum content of toxic substances in industrial waste for determining toxicity levels (UVT. AS USSR 27.12.1984, Ministry of Health, USSR, 18 December 1984 N 3170-84).
18 Decree of the President Russian Federation from 13 May 2000, N 849, "About Authorized Representative of Russian Federation President in Federal District."
19 Amendments to the Federal Law "About Waste of Production and Consumption" and Selected Law Acts of the Russian Federation, # 396708-5, http://www.eg-online.ru/document/law/105092/ (accessed 29 November 2011).
20 H. A. Djuvelikian, "Ecological Condition of Virgin and Anthropogenic Landscapes in Central Black Earth Zones," Dissertation abstract, Petrozavodsk State University, 2007.
21 M. P. Lobanova and T. A. Trofimova, "Heavy Metal Pollution on Territory of Volgograd Region," *Proceedings from the Lower Volga Agricultural University Complex*, 1(9), 2008, 12–16.
22 K. A. Pasika, "Investigation of Cement Dust Emission Influence on Growth and Evolution of Plants," *Advancement of Modern Natural Science* 11, 2004, 45.
23 M. V. Kudin, D. E. Tsimbal, A. V. Skripkin, and I. N. Federov, "The Ecology of the Environment in a Region with developed cement industry, " *Problemy i perspektivy sovremennoi meditsiny, biologii i ekologii* 1,4, 2010, 33–36.
24 K. A. Cherepanov et al., "Utilization of Secondary Material Resources."

25 O. S. Kolbasov, "Environmental Law Administration and Policy in the USSR," *Pace Environmental Law Review* 5/2, 1988, 439–44.

26 M. V. Kudin, "Ecogeochemical Characteristics of Regions with Developed Cement Industry," *Saratovskii Nauchno-Meditsinskii Zhurnal* 7/1, 2011, 26–30.

27 G. G. Volokitin, N. K. Skripnikova, N. A. Pozdniakova, O. G. Volokitin, and A. V. Lysenko, "High Temperature Methods for Cement Clinker Manufacturing Using Low Temperature Plasma and Electric-Arc-Driven Heating (Joule Heating)," *Tomsk State Architectural-Construction University Herald* 4, 2008, 106–13.

28 J. Lafond, R. R. Simard, and D. Pageau, "Agronomic Evaluation of Cement Kiln Dust in Potato Production," *Canadian Journal of Soil Science* 73 (4), 1993, 645.

29 D. V. Bolshakov and E. G. Pikhteeva, "Comparative Evaluation Metabolic Impairment when Exposed to Small Amounts of Cadmium and Mercury," *Aktual'nye Problemy Transportnoi Meditsiny,* 1(3), 2006. 12–18.

30 Ibid.

31 Ibid.

32 I. A. Pereslegina and P. P. Zagoskin, "Biochemical Characteristic of Interaction between Level of Lead in Children's Blood and Damage of Central Nervous System Function," *Nizhegorodskii Meditsinskii Zhurnal*, 8, 2006, 139–48.

33 I. M. Zhannikov, T. G. Gabrichidze and L. T. Zubko, "Investigation of the Influence of Arsenic-Rich Compounds and Possibility of Predicting the Critical Situation on Chemical Dangerous Object," *Intellectual Systems in Industry* 9, 2007, 113–18.

34 A. G. Pirchelani, N. A. Pirchelani, R. A. Gakhokidze, N. V. Bichishvili and E. A. Kalandia, "Effect of Poulsen Vitamin Complex on Mutagenic and Cytotoxic Influence of $CuCl_2$ in Experiment," *Georgian Medical News* 6 (159), 2008, 44–46.

35 Bolshakov and Pikhteeva, "The Comparative Evaluation of the Metabolic Impairment"; Reilly, *Metal Pollution of Food*.

36 H. M. Ovsiannikova, "The Specifics of Adaptation Reaction of People with Heavy Metals in Their Organisms," *Memoir of Tavrichesky National University of V. I. Vernadsky*, 23(62), 2010, 142–51.

37 Cherepanov et al., "Utilization of Secondary Material Resources."

38 Ibid.

39 O. L. Tavrovskaia, "About the Utilization of Metal Industry Waste," *Agrokhimiia* 4, 1992, 55–61.

7 Environmental contamination and public health in the Republics of Tatarstan and Mari El, Russian Federation

Nailya Davletova

The state of public health is influenced by many environmental phenomena, including air, drinking water, food, and soil. It is well known that in recent years, demographic indicators have worsened and morbidity has increased in the Russian Federation. One of the likely reasons for this trend is the environmental distress in many regions, especially in industrial cities. Researchers have estimated that 20 percent of any given population's health is affected by the environment.[1] Evaluating the quality of the habitat and the relationships between environmental components and health are important prerequisites to adopting measures aimed at reducing the risk of contamination in a timely fashion.

The data used in this research reveal the importance of the quality of the environment, the presence of significant industrial footprints on certain areas, and the influence of pollution on human health. The study of the environment and its impact on health in the Volga Federal District began in the mid-1980s, with the rapid development of industry and accumulation of petroleum deposits in Tatarstan. However, the earlier studies were local in nature and examined the individual components of the environment (air, water, soil) and the impact of pollution of those individual components on public health in these areas. There was also no research that would have addressed the question of the relationship between money spent on environmental protection and the ecological situation in some areas. Therefore, despite the large number of studies on the adverse effects of the environmental factors, there is still a need for quantitative evaluations of their contribution to the formation of long-term consequences for public health.

According to research by F. F. Dautov,[2] V. P. Rassanov,[3] L. Z. Rashitov,[4] L A. Gabdullina,[5] A. P. Rossolovskii,[6] and N. N. Mitina,[7] the influence on the population's health is broadly distributed. Sources of contamination are one of the anthropogenic indicators in a given populated area. In some districts we observe a significant industrial and this in turn leads to the increase in the populations' morbidity in respiratory, cardio-vascular, urinary, and gastro-intestinal diseases.

By studying the quality of the environment through a rigorous evaluation of the contaminants and risk factors, it is possible to design preventive actions that will improve the environment and public health. Moreover, on the basis of the analysis of the effect of environmental factors on public health, we are able to determine which clean-up measures have the greatest urgency. An analysis of the influence of environmental factors on public health helps to prioritize responses and implement preventive measures for selected locations. This saves time and money for state and private decision-makers of enterprises and other officials charged with protecting the environment and human health. It is also important to optimize the ecological and hygienic security of the population and justify the sequence in which preventive measures are implemented in individual residential areas.

The goal of this study is twofold. First, the author undertakes a multifaceted environmental-health evaluation of environmental pollution. Second, she assesses the influence of pollution on public health in order to develop preventive measures. The research goals are to:

1   perform an ecologic analysis using publicly available data sources on regional pollution and health indicators to determine areas of greatest need/greatest risk of adverse health outcomes;
2   analyze the regional characteristics of environmental components and their contribution to environmental degradation in Tatarstan and Mari El;
3   study the health of populations in Tatarstan and Mari El;
4   identify and rank territories with dangerous levels of pollution as a result of industrial processes according to the criteria of environmental footprint, quality of air and water, and public health;
5   develop recommendations for implementing preventive policies for improving the environment.

Methodology

The research was conducted in the Russian territories of Tatarstan and Mari El in the Volga region of Russia and focused on the three main components of the environment (air, drinking water, soil) and their impact on the health of the population in these republics. The study was conducted in four stages. At every research stage the corresponding research objective, the subject of research, sources, methods of information gathering, processing and analysis were identified. The study employed a hygienic method (in part a method of sanitary expertise that compares the indicators of contamination with normative indications). In addition to the hygienic, epidemiological, geographical, sociological, mapping, analytical, and statistical methods were also applied. This research was reviewed by experts in the medical and environmental fields and expert surveys were conducted to identify the most important indicators and weight coefficients for each factor in the total load.

The data used for this analysis were taken from official open sources, presented in the form of government reports, official forms of state statistical reports, and the database of "Key Indicators of Municipalities" of the Territorial Agency of the Federal State Statistics Service of Tatarstan and Mari El. One of the challenges in

Table 7.1 Stages and scope of research

Research stages	Research methods	Information sources
Literature review and analysis	Descriptive analysis of exposure and outcome data sources. Analytical method	Literature review of Russian and foreign authors (108 sources)
The analysis of regional differences in quality of components of environmental health (air, drinking water, food, soil, etc) and their impact on environmental degradation of the territory.	Hygienic method. Peer review. Statistical analysis. Evaluation of air pollution. Evaluation of the degree of air contamination according to indicator "R" and the Aggregate Index of Air Pollution	Analysis of the materials of the Center of Meteorology and Environmental Monitoring, The Center for Monitoring of Water Resources, reports on air and water, according to materials of the State Report on the Environment
Examination of public health using morbidity data	Epidemiological, statistical analysis based on the research findings of the F.F. Erisman Federal Scientific Centre of Hygiene "Assessment of the epidemiological health risks at the population level in medical and hygienic ranking areas"*	Official state statistical data and research for 2003–2009
Identification of degraded territories by ranking of criteria of industrial footprints and indicators of the quality of the components of the ecological system and population health.	Statistical method, mapping	Research results
Crafting preventive recommendations that will be implemented to improve the quality of the environment and of public health.	Analytical method	Research results

*A. I. Potapov, "Assessment of epidemiological health risk at the population level in the medical-hygienic ranking of territories" in A. I. Potapov et al., *Sbornik metodicheskikh rekomendatsii zdorov'e naselenie i okruzhaiushchaia sreda*, No. 3. Part 1. Moscow: Federal Center of State Sanitary-Epidemiological Monitoring, Ministry of Health of Russia. 1999, 494–503.

collecting data was the generalized nature of the data describing the expenditures on environmental protection; there was little to no information about funding sources. Therefore, this section presents an analysis of the amount of funds spent on environmental protection by districts of Tatarstan and Mari El. The stages, methods, and research sources are presented in Table 7.1.

The environment and public health

A favorable environment is an environment that sustains the natural ecological systems and anthropogenic objects.[8] In recent years, there has been a marked deterioration in demographic indicators in the Russian Federation, especially rising morbidity and reduced life expectancy. One of the reasons for the poor demographic trends is the environmental tension between manmade pollutants and nature.[9] Numerous studies of the factors influencing public health conducted in the Russian Federation and abroad clearly show that pollution has a negative impact on human health.[10] For example, according to E. V. Baklazhenko, the pregnancies of women living in conditions of environmental degradation are characterized by a decrease in blood suppressors or cytoxic cells, by a disturbance in the balance of immune-regulating subpopulations and a reduction in the production of immunoglobulin. The negative impact of pollution on the dynamics of nephrological pathology and respiratory disease has also been studied.[11] Indeed, there are many suspected links between environmental contamination and adverse health outcomes: air pollution and asthma/cardiovascular disease/death; industrial chemicals (occupational exposure) and cancer; and pesticide use and male and female reproduction.

The pollution of the environment, causing the degradation of the human habitat and bringing harm to people's health, is one of the most pressing environmental problems facing the Russian Federation today, in social and economic realms in addition to public health.[12] Distinctive environmental problems in regions of the Russian Federation stem primarily from industrial, transportation, utilities, and agricultural pollutants. The intensity of pollution depends on the industrial or agricultural plants' proximity to population centers as well as their technological level.[13] Although there are a large number of studies on the adverse effects of environmental factors on human health, few quantitative assessments of their long-term consequences for human health have been undertaken. In addition, an integral assessment of the industrial footprint and geo-chemical factors, and their impact on the imbalances between urban and rural populations, are still understudied topics.

The contribution of manmade pollutants in forming deviations in health indicators varies from 10 to 30 percent. Most authors emphasize the importance of monitoring air, drinking water, and sediments, including soil and snow cover.[14] To ensure that vital economic activity in both urban and rural regions does not adversely health requires a complex assessment of the interplay of environmental and social ecosystems.[15]

Major manmade pollutants violate the natural biochemical processes in the human population and this gives rise to the development of pathologies.[16] The assessment and prediction of negative consequences on the public's health status are taken into account in studying the role of independent factors of the environment. In assessing and predicting the role of water on public health, areas are ranked by the severity of the problem in connection with the implementation of economic and environmental programs.[17] Public health suffers from lower than acceptable hygienic norms (for air pollution, the appearance of unusual samples of drinking water, soil) that in turn cause functional disturbances in the central nervous system, metabolic activity, cardiovascular, respiratory and digestive systems, and lead to increased morbidity and mortality.[18]

The pollution problems of the cities, especially the largest, are characterized by the excessive concentration of transportation and industrial enterprises and the formation of manmade environments that are ecologically unbalanced.[19] Air pollution poses a serious threat to public health, and contributes to the decline of quality of life.[20] In recent years, trends in the growth of cancer, leukemia, and other life-threatening diseases have been observed.[21]

Water and soil pollution

The purity of drinking water in cities and rural areas is extremely important. The provision of potable drinking water to the population is one of the main indicators of quality of life of the population.[22] Indeed, the quality of drinking water, its chemical composition, and absence of disease pathogens play a significant role in maintaining public health.[23] It is a paradoxical situation in Russia, the country possesses huge reserves of natural water, yet is unable to provide uncontaminated water to satisfy its drinking and domestic requirements.[24] The main reasons for such an unfortunate situation are: grave pollution of water sources, old and outdated water purification technologies, and violations of technical standards of existing purification facilities. In addition, secondary water pollution from water-transfer networks due to the low quality and corrosion of the pipes, inadequate sanitary-technical levels of construction and maintenance, and significant pressure drops in transfer networks in some residential areas are also problematic.[25]

Soil absorbs chemical substances of manmade origin and thereby adversely affects the lives and health of the population, but organic materials in soil can also act as a purification system, removing unwanted environmental contaminants. Heavy metals are among the most common types of anthropogenic pollutants found in the soil.[26] Soil contamination from heavy metals represents a significant portion of chemical pollutants that puts human health at risk.[27] Many heavy metals are highly toxic in trace amounts. They also have significant migration ability and propensity to bio-accumulate. Chemicals penetrate the human body at concentrations greater than that found in the environment,[28] which makes their presence, even in small concentrations, dangerous to human health.[29]

To address important regional problems, integral assessments of risk factors in urban areas, identification of top priorities, establishment of informative health indices dependent on the ecological situation, and the development of regional prevention campaigns must be undertaken.[30] The protection of nature and public health are closely interwoven in modern cities where a variety of large and small industrial and energy utilities and household sources of physical, chemical, and biological pollution are concentrated.[31] Automobile transport and the growth of the number and power of the vehicles play a large role in polluting the urban environment. Indeed, large cities are becoming the epicenters of polluted air, water, and soil and endangering the sanitary and epidemiological welfare of the population.

The development and implementation of preventive measures to protect and improve public health and the analysis and identification of areas where the environment adversely affects public health are vital. It is essential to take an interdisciplinary approach to the creation of practical mechanisms to prevent morbidity caused by pollution at local and regional levels based on hygienic diagnostics of pollution's effect on health.

Regional disparities in environmental components and their impact on hygiene and public health in Tatarstan and Mari El

The analysis of the environment in Tatarstan and Mari El was conducted on the basis of many documents, including environmental monitoring data, records of federal and regional executive agencies responsible for state environmental control and natural resource management, and data from state reports on the environment. These data were used in analyzing the regional differences in environmental quality components and their contribution to the environmental and hygienic degradation of Tatarstan and Mari El.[32]

The main characteristics of the republics are presented in Table 7.2.

Air pollution

The atmosphere is an essential component of the ecological system and it has significant impact not only on the hydro-and lithosphere, but also on human health. Air in the cities and towns of Tatarstan and Mari El is contaminated by natural sources, such as soil dust, dirt from rotting of organic materials released from wetlands, forest fires, industrial emissions, as well as emissions from municipal facilities, including bus, car and train transportation and trans-boundary pollutants. The latter is most typical of Mari El, where air pollution from the surrounding territories of Kazan, Zelenodolsk, Cheboksary, Novocheboksarsk, and Nizhny Novgorod contaminate the republic. Every year more than one million tons of pollutants are carried by the wind to Mari El.

The main source of air pollution in residential areas of Tatarstan and Mari El is auto transport. Despite the decline in total emissions from stationary sources (such as industrial enterprises) the environment has not improved because of automobile pollution.

However, unlike Tatarstan, the monthly average concentrations of all monitored components in the cities of Mari El for the studied period did not exceed acceptable standards. This is a result of the shape of the terrain and the climatic conditions favorable for the diffusion of pollutants in Mari El. These features render Mari El a republic with a lesser propensity to air pollution.

Air emissions from stationary sources in Tatarstan emanate from the oil and gas industry (32.6 percent), heat and power complex (32.7 percent), chemical and petrochemical industries (19.0 percent), machine building (5.3 percent), and construction (2.7 percent). Remaining enterprises contribute 7.7 percent of total emissions into the atmosphere

The leading pollutants include hydrocarbons, such as light volatile organic compounds, nitrogen oxides, sulfur dioxide, carbon monoxide, and suspended solids.

Water pollution

The reserves of safe drinking water in Mari El are estimated at 3.2 million cubic metres per day. According to a government report on the environmental situation in Mari El, typical water pollutants are compounds of iron, manganese, copper, and easily oxidized organic matter: nitrites, phenols, and oil products. The majority of pollutants enter the river basins in areas with a high concentration of industries, housing, and communal services, which are major polluters of water. Most of the water bodies fall into the third group of moderately polluted waters. Pulp and paper industries, as well as treatment facilities of Yoshkar-Ola, contribute to more than 30 percent of water pollution.

The proportion of groundwater in the overall balance of domestic water supply in Mari El is 86–100 percent, and surface water used for drinking water is found only in the cities of Yoshkar-Ola (14 percent) and the Volzhsk (8 percent), usually mixed with groundwater. The main waste-producing enterprises of the republic are the Mariskii Pulp and Paper Company, (bark waste, sawdust, waste cellulose fiber), the Fokinskii Distillery (alcohol grains), mechanical and biological sewage treatment facilities, and agricultural and agro-processing companies (dung, manure, waste from cleaning of grain, whey).

According to data from Tatarstan's Ministry of Ecology and Natural Resources, the water quality of small rivers in Tatarstan is characterized as "moderately polluted" and "dirty." Contamination of small rivers is caused by diffuse sources and the level of pollution depends on the meteorological conditions. Ground waters in Tatarstan are characterized by increased (from 7.03 to 42.4 mmol/dm^3) values of total hardness, elevated concentrations of iron (from 0.31 to 10.6 mg/dm^3), and sulfates (from 519.2 to 1793.4 mg/dm^3). The reason for elevated levels of water hardness is the presence of water-bearing rocks, gypsum, and anhydrite, pulling tainted water from underlying aquifers.

Hygienic ranking is a method used to analyze the sanitary-epidemiological situation in the area. The results of the analysis are applied to the development of the management system of preventive activity. The methodology of calculation and the basis on which areas were ranked are presented in Appendix 1. The territories of the republics were divided into groups. The first group included the regions where an integrated "environmental indicator of poor health" (EIOPH) was higher than the average. These are so-called "degraded areas according to EIOPH." The second group consisted of areas with an EIOPH lower than the average for the country and thereby considered "safe areas" according to EIOPH.

After identifying the areas with a high EIOPH index, an analysis of the contribution of individual environmental components to this index was performed. The ranking results are presented in Appendices 3 and 4.

Table 7.2 Geographical characteristics of the Republics of Tatarstan and Mari El

	The Republic of Tatarstan	The Republic of Mari El
Geographical location	Located on the East-European Plain at the confluence of two major rivers – the Volga and the Kama	Zone where the Volga Uplands meet the East European Plain
Borders	To the west – Republic of Chuvashiia; to the east – Republic of Bashkortostan; to the north-west – Republic of Mari El; to the north – Republic of Udmurtiia and Kirov Oblast; to the south – Orenburg, Samara and Ulyanovskaia Oblast	To the west – Nizhegorodskaia Oblast; to the north-east – Kirovskaia oblast; to the south-west – Republic of Tatarstan; to the south – Republic of Chuvashiia
Area, length	Area: 67,836 km2; length north to south: 290 km; length east to west 460 km	The Area : 23,400 km2; length north to south 150 km; length from east to west 320 km
Territorial division	43 municipal districts and 2 urban districts (Kazan and Naberezhnye Chelny)	14 districts, 3 cities (Yoshkar-Ola, Volzhsk and Kozmodemiansk)
Population	Total: 3,780,600; urban: 2,831,670; rural: 948,930	Total: 711,500; urban: 449,100; rural: 262,100
Climate	Temperate continental	Temperate continental
Topography	Plain in the forest and forest-steppe zone with small hills on the right bank of the Volga and the south-east of the country; 90% of the territory lies at an altitude of less than 200 m above sea level	Rolling plain
Hydrography	Major rivers: the Volga and the Kama and two tributaries of the Kama – Viatka and Belaia, about 500 smaller rivers	Major rivers: the Volga and the Vetluga; about 500 large and small rivers
Soil	Soils are very diverse – from the gray forest and ashen-gray (podsolic) in the north and west to the different types of black soil in the south (32% of the area). There are particularly fertile black earth soils in the region, but gray forest soil and leached black soils dominate	In the southern part -sandy, sandy-loamy; in the north – clay, loamy; in low-lying parts – peaty

Key industries	Oil and gas exploration and drilling, chemical, petrochemical, machine building, light and food industries	Engineering and metalworking for production of refrigeration equipment, semiconductors, machinery for forestry, road construction equipment, components and assemblies for excavators, artificial leather
Agriculture	Agricultural land in farms of all categories: 4152 hectares (61%); arable land: 3413 ha (50%). Products include grain, sugar beet, potatoes, poultry, meat and dairy products as well as livestock areas, horse breeding and beekeeping.	Agricultural land occupies 35% of the republic. Products include rye, buckwheat, peas, oats, wheat, flax and potatoes, as well as cattle, pigs, sheep and goats
Distribution of land	Agricultural land 4,662,700 ha; forest 1,241,400 ha; water resources 438,700 ha; other land 2,500 ha	Agricultural land 786,800 ha; surface water 84,800 ha; marsh 34,700 ha; forests and woods 1,346,900 ha; other land 84,300 ha

Note: Based on the official website of the Republic of Tatarstan (www.tatar.ru) and the official website of the Government of the Republic of Mari El (www.gov.mari.ru), and a social atlas of Russia

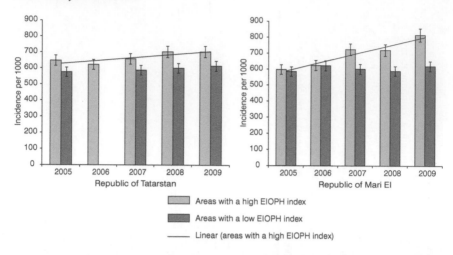

Figure 7.1 Morbidity in Tatarstan and Mari El in areas with different levels of EIOPH index, 2005–2009

Figure 7.1 shows the average regional values for EIOPH was 9.45 Tatarstan and 8.4 for Mari El.

According to the research findings, the major contributor to the EIOPH index is the quality of drinking water: its contribution to the integral index varies from 52.71 percent to 30.46 percent for the districts of Tatarstan and from 43.43 percent to 26.61 percent for Mari El. The contribution of air pollution was from 37.66 percent to 12.12 percent for the regions of Tatarstan and from 33.29 to 17.97 percent for the regions of Mari El. The contribution of soil contamination to the index was 26.63 to 9.68 percent for the regions of Tatarstan and 27.17 to 17.92 percent for the regions of Mari El. The impact of nature, climate conditions, and food quality is small and averages 11.91 percent (Tatarstan), 12.23 percent (Mari El) and 7.73 percent (Tatarstan), and 7.84 percent (Mari El) respectively.

Environmental protection measures are evolving in the two republics. The cost of environmental protection activities in 2009 averaged 220,225,700 rubles and 27,922,790 rubles (or US$7,104,055 in Tatarstan and US$900,735 in Mari El). The amount of funds spent on nature protection in 2009 is presented in Table 7.3 for districts with high values of EIOPH.

As is evident from Table 7.3, approximately US$6,849,359 are spent annually on environmental protection. In some cases, the funds allocated are not adequate to protect the environment on the territory of a given municipal district.

For example, the Sarmanovskii and Drozhanovskii districts of Tatarstan are agricultural districts with similar environmental conditions, but the amount of money spent on environmental protection in Sarmanovskii districts is larger, and equal to US$12,678,177, while in Drozhanovskii district it was US$52,367. At the same time, in Zelenodolskii district with well-developed industrial sector and thus with more sources of pollution, (which was confirmed by the high rate of EIOPH) the amount of funds spent on environmental protection is only US$4,514,106. In

Table 7.3 2009 public spending on environmental protection of districts in the Republics of Tatarstan and Mari El with high levels of contaminants

Districts	The value of EIOPH index	Expenditures (US$*)
Republic of Tatarstan		
Menzelinskii	9.48	24,296
Vysokogorskii	9.54	82,551
Sarmanovskii	9.58	12,678,177
Drozhzhanovskii	9.59	52,367
Cheremshanskii	9.72	29,638
Agryzskii	9.79	56,329
Almet'evskii	9.84	59,418,516
Verkhneuslonskii	9.88	172,271
Buinskii	10.05	148,032
Baltasinskii	10.11	65,113
Elabuzhskii	10.14	7,289,787
Mendeleevskii	10.18	1,629,961
Atninskii	10.19	n/a
Chistopolskii	10.24	559,264
Pestrechinskii	10.80	205,629
Zainskii	10.82	1,489,258
Nurlatskii	10.93	5,879,832
Zelenodol'skii	10.96	4,514,106
Tukaevskii	11.33	29,894,116
Nizhnekamskii	11.60	68,457,290
Kazan	12.32	45,147,387
Republic of Mari El		
Medvedevskii	8.41	1,593,879
Novotor"ial'skii	8.58	312,795
Sernul'skii	8.58	329,370
Gornomariiskii	8.86	26,654
Yoshkar-Ola	9.51	6,058,448
Zvenigovskii	9.67	787,583
Koz'modem'iansk	9.69	1,477,409
Volzhsk	10.84	2,174,564

Note. Based on the Regional Departments of the Federal State Statistics Service for Tatarstan and Mari El: data for Mari El is from http://maristat.mari.ru/scripts/munsInet/DBInet.cgi; data for Tatarstan is from http://www.gks.ru/dbscripts/munst92/DBInet.cgi; data for Russian Federation is from http://www.gks.ru/dbscripts/munst/munst.htm
* For the purposes of this calculation, US$1 = 31 rubles.

Table 7.4 Distribution of the population according to air and water quality

Air/water quality groups*	Republic of Tatarstan		Republic of Mari El	
	Population**	Percentage	Population**	Percentage
Air quality				
Group 1	2,570,343	68.84	416,900	58.39
Group 2	329,768	8.83	80,200	11.23
Group 3	333,103	8.92	114,600	16.05
Group 4	500,425	13.41	102,300	14.33
Quality of drinking water				
Group 1	652,327	17.5	143,100	20.04
Group 2	1,593,908	42.7	72,600	10.17
Group 3	1,172,125	31.4	388,400	54.40
Group 4	315,279	8.4	109,900	15.39

*   where the quality of the air/drinking water for Group 1 can be described as "extremely unsatisfactory", for Group 2 as "satisfactory", for Group 3 as "acceptable" and for Group 4 as "good."
**   based on the Regional Departments of the Federal State Statistics Service for Tatarstan and Mari El.

this case it is necessary to identify the top-priority needs in order to optimize the allocation of financial resources. Roughly 47 percent of districts of Tatarstan and Mari El have a higher index of environmental and health distress, especially in the areas of air and drinking water quality. The territory of Tatarstan and Mari El were ranked by the quality of drinking water and air quality. All areas were divided into four groups (characteristics of the groups are presented in Appendix 4). The ranking identified the territories where the quality of the water/air was lower and contribution of the quality of drinking water/air into the EIOPH is higher than the average for the republic.

The hygienic ranking in Tatarstan and Mari El, based on the ecological and hygienic distress rate and EIOPH, revealed twelve critical areas of poor quality drinking water districts in Tatarstan and three districts and one city in Mari El. As for air pollution, twelve districts of Tatarstan, including Kazan, and one district and three cities in Mari El have the highest concentrations of pollution (see Table 7.4).

Table 7.4 illustrates that 17.5 percent of the population of Tatarstan and 20.04 percent of the population of Mari El live in areas with poor quality drinking water. Similarly, 68.84 percent of the population of Tatarstan and 58.39 percent of the population of Mari El live in areas of poor air quality. These territories include the major cities and towns with well-developed transportation networks and industries. Based on the research data it is possible to assume a rise in overall morbidity as well as clinical forms of morbidity, as they are common in conditions of polluted air and contaminated drinking water.

Public health in Tatarstan and Mari El

Human health is partly determined by the environment, which consists of natural phenomena (the environment, flora, landscape features, etc.) and anthropogenic or "manmade" phenomena, such as chemicals that pollute the air, soil, water; noise, and electromagnetic fields. The quality of the environment is one of the main determinants of public health. According to the World Health Organization (WHO), 20 percent of human health is influenced by the physical environment.[33] To analyze the influence of environmental pollution on adverse public health outcomes, the author calculated relative probabilities of health risks of diseases in two highest exposure regions using group-level exposure data. The probability of epidemiological risk is the ratio of unfavorable health-demographic indicators (morbidity, mortality) in studied regions to baseline rates. Figure 7.1 shows the relative indicators of risks were calculated for 45 districts of Tatarstan between the years 2003 and 2007.

The analysis in Figure 7.2 shows that Tatarstan has a variety of relative risks. There is an increase in the areas in which the relative risks of morbidity are estimated as high (three-fold increase in 2003: up to 10 times in 2007). At the same time there are more regions with unacceptable morbidity risks. This indicates a possible increase in morbidity among the population of the republic of Tatarstan. The results of the analysis of morbidity data for Tatarstan and Mari El (calculated as the number of people newly diagnosed with disease), according to the regions with different levels of EIOPH index, are shown in Figure 7.1.

As seen from Figure 7.1, territorial morbidity levels differ according to levels of EIOPH. For instance, the morbidity rates in districts of Tatarstan and the quality of the air/drinking water for Group 1 can be described as "extremely unsatisfactory," for Group 2 – as "satisfactory," for Group 3 "acceptable," and for Group 4 as "good." EIOPH index are higher on average by 11.65 percent than

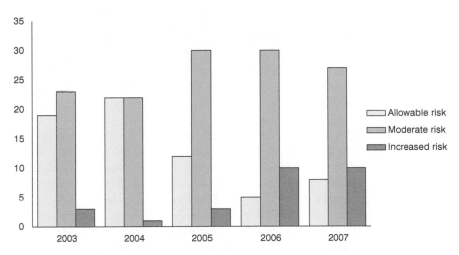

Figure 7.2 Relative risks of disease in relative industrial areas in Tatarstan, 2003 to 2007

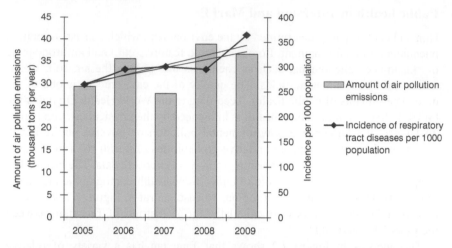

Figure 7.3 Respiratory diseases in the Republic of Mari El and air pollutants from stationary sources, 2005 to 2009

those in areas where the EIOPH rate is below the average level for Tatarstan. For Mari El the difference in morbidity is 11.67 percent ($p <0.05$).

Thus, higher rates of morbidity are found in areas of high ecological and hygienic distress index, as compared with the territories where this index is lower than the average regional rate.

Figure 7.3 shows the analysis of air quality and incidence of respiratory diseases of the Republic of Mari El shows the following: during 2005–2009, emissions of pollutants into the air from stationary sources increased by 17.97 percent and the incidence of respiratory diseases of the population by 27.56 percent.

The analysis suggests that the high incidence of certain types of abnormalities is related to the quality of water consumed, i.e., disease risks are related to environmental exposures. Therefore, an evaluation of the reliability of differences in levels of morbidity (Class of ICD-X), affecting the population residing in the research areas was undertaken. The levels of morbidity from cardiovascular, urogenital, digestive diseases and diseases of blood and blood-forming organs, (such as those classes of diseases that may result from the quality of drinking water), were examined. Figure 7.4 shows groups of the districts of Tatarstan ranked according to the EIOPH index.

Figure 7.4 shows classes of diseases according to the World Health Organization's Classification of Disease: III – diseases of blood and blood-forming organs; IX – cardiovascular diseases; XI – diseases of the digestive system; XII – diseases of the skin and subcutaneous tissue; XIII – diseases of the musculoskeletal system; XIV – diseases of the urogenital system.[1]

A comparative analysis in Figure 7.5 shows that in areas with poor-quality drinking water, the level of cardiovascular morbidity increases by 23.4 percent, of musculoskeletal system by 32.2 percent and urinary tract by 31.3 percent than in areas where the population has access to water of higher quality ($p <0.05$).

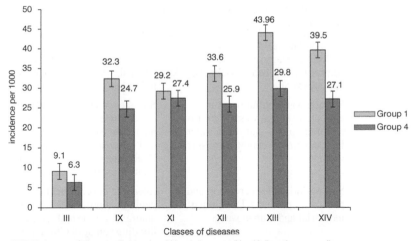

MKB-X classes of disease: III, blood and blood diseases; IX ... XI digestive organ diseases; XI skin and tissue disease; XIII musculoskeletal diseases, and XIV urogenital diseases.

Figure 7.4 Morbidity levels in Tatarstan as a function of drinking water quality

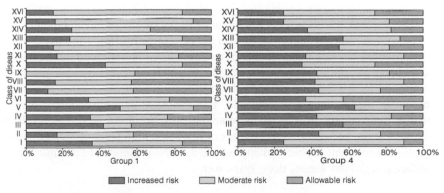

Figure 7.5 Relative epidemiological risks of diseases in the Republic of Tatarstan in the districts of Groups 1 and 4 as a function of quality of water consumed

As can be seen from Figure 7.5, drinking water quality is one of the key influences on the health of the population. For instance, in the regions of Group 1 with low quality drinking water, the percentage of areas with increased relative risk of diseases is above average. All clinical forms of moderate risk are observed. At the same time, Group 4 regions with an acceptable risk of diseases dominate.

Thus, the composition of drinking water influences the risk of diseases in general. Residents who consume drinking water of poor quality, unbalanced in microelements and salt content, are at risk of developing infectious disease. When the situation worsens, the failure of adaptive mechanisms and the development of disease in some individuals are likely to develop.

Table 7.5   Key environmental problems in the Republics of Tatarstan and Mari El

Element	Republic of Tatarstan	Republic of Mari El
Air	High and unhealthy levels of air pollution in large cities (Kazan, Naberezhnye Chelny, Nizhnekamsk, Almetevsk)	High and unhealthy levels of air pollution in cities of Yoshkar-Ola and Volzsk. Cross border transfer of pollutants
Water	Surface water pollution from untreated sewage wastewater from treatment plants and storm water. Hydrochemical condition of surface water	Surface water pollution by untreated sewage wastewater from treatment plants and storm water. Flooding and water-logging in zones of the Cheboksary and the Kuibyshev Water Reservoirs
Soil	Soil pollution from crude oil and petroleum products. Dumping of industrial and human waste	Intensive development of exogenic geological processes (landslides, landfalls, ravine and river erosion, flooding) on agricultural and residential lands

Sources: Government Report " On the state of environment in the Republic of Tatarstan for 2005–2010," http://eco.tatar.ru/rus/info.php?id=48562; The Report on State of Environment in Mari El for 2010, http://www.gov.mari.ru/debzn/s_ekolog.shtml

The research has shown the relevance of issues of environmental quality and revealed a significant industrial burden on certain territories. Moreover, it has demonstrated the significant impact of environmental pollution on public health. Anthropogenic pressure on the environment is demonstrated by changes in air quality, surface water and groundwater, and soil. In contrast to other regions of the Volga region, including Tatarstan, the manmade or anthropogenic burden on the environment in Mari El is relatively low, resulting in minor levels of contamination. However, in some areas of Mari El a high level of human impact on the environment is observed and reflected in the presence of contaminants in surface and ground waters, flora, soil cover.

Conclusions

In conclusion, pollution is largely a local phenomenon. To prevent the build-up of pollutants, which can spread to large areas and be irreversible, constraints must be placed on the development of environmentally hazardous industries. The environmental and economic interests of the territory must be coordinated with the introduction of environmentally safe technologies and efficient use of natural resources. Environmental stability may also be achieved through the targeted use of federal and republic budgets for ecological activities and through the investment in the cleanest, greenest available technologies for industrial production.

Specific conclusions derived from the study include:

1   The Republics of Tatarstan and Mari El suffer from poor quality natural groundwater (increased rates of total hardness, elevated concentrations of

iron, sulfate); poor quality of surface waters due to anthropogenic pollution; local air pollution in major cities with developed infrastructure and industry. In addition, for the Republic of Mari El, the cross border movement of pollutants from neighboring republics is still a pressing problem.

2   The main contributors to the EIOPH index in the studied areas are the quality of atmospheric air (contributes from 37.66 to 12.12 percent in regions of Tatarstan and from 33.29 to 17.97 percent in regions of Mari El) and the quality of drinking water (contributes from 52.71 to 30.46 percent in the regions of Tatarstan and from 43.43 to 26.61 percent in Mari El).

3   Public health in the Republic of Tatarstan and Republic of Mari El is characterized by increased morbidity in areas with high values of EIOPH. For example, morbidity in the districts of Tatarstan with a high EIOPH index is on average 11.67 higher than areas where EIOPH index is below the republic average (p <0.05). A comparative analysis shows that in areas with contaminated drinking water the level of morbidity for cardiovascular diseases is significantly higher by 23.4 percent; of diseases of the musculoskeletal system, by 32.2 percent and urogenital diseases by 31.3 percent than in areas where the population has access to water of higher quality (p <0.05). A weak positive correlation between the amount of pollutants in the air from stationary sources and morbidity of the population of respiratory diseases was found.

4   Each year approximately 336,031,900 rubles is allocated to the protection of the environment by the Russian Federal Government. In some cases, the funds allocated are inadequate for the environmental problems in a given municipal district. It is imperative to identify and rectify polluted areas and make better use of financial resources dedicated to environmental management.

5   As a result of this study, twelve districts in Tatarstan (Agryzskii, Atninskii, Baltasinskii, Bugulminskii, Vysokogorskii, Zelenodolskii, Mendeleevskii, Musliumovskii, Nurlatskii, Pestrechinskii, Cheremshanskii, Chistopolskii municipal districts) and three districts and a city in Mari El (Zvenigovskii, Novotorialskii, Sernurskii municipal districts and Volzsk) were identified as having poor quality drinking water. The elevated degree of air pollution in twelve districts of Tatarstan, including Kazan, and one municipal district and three cities in Mari El render them unhealthful as well. Algorithms of measures to optimize the quality of drinking water and air in order to minimize their impact on public health have been developed.

Practical recommendations

Measures to optimize the quality of environmental components should be based on a comprehensive analysis of the ecological-hygienic situation in a particular area, which in turn should include analysis of the major components of the environment (air, climatic conditions, water quality and soil) as well as analysis of population health. Managerial decisions to optimize the quality of the environment should be made on the basis of such data. As a result of this research preventative measures to improve the environment have been provided in Figure 7.6.

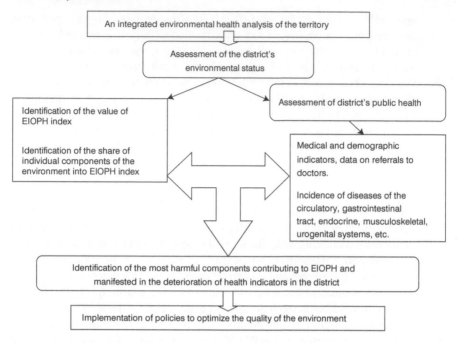

Figure 7.6 Algorithm for improving the environment in areas with the worst pollution

The degree of air pollution in 12 districts of Tatarstan (Almetievskii, Elabujskii, Zainskii, Zelenodolskii, Mendeleevskii, Tukaevskii, Nizhnekamskii, Nurlatskii, Cheremshanskii, Chistopolskii municipal districts), and city of Kazan, and one municipal district and three cities in Mari El (Medvedevskii municipal district, cities of Yoshkar-Ola, Vol'zsk, Kozmodemiansk), can be considered the most problematic. The algorithm of measures to optimize the quality of air in order to minimize the impact on health outcomes in these areas can be shown as follows:

Although measures to protect air and soil are usually long term and capital intensive, providing the population with high-quality drinking water as quickly as possible is a top priority. In Tatarstan, the problem is providing adequate amounts of drinking water to serve 12 municipalities in the area inhabited by 652,327 people (17.5 percent of the population), for drinkable water that will require 1,956,981 litres/day. Table 7.6 shows that to provide drinking water to 143,100 residents of distressed areas of Mari El (20.04 percent of the population) will require 423,300 litres/day.

The fastest, most reliable and radical way to ensure the supply of safe drinking water is to use water packaged in containers as an alternative drinking water source. As the first stage in optimizing the water supply in Mari El, this will bring quick results without high economic costs. The possibility of organizing the production of bottled water using wastewater treatment plants should be also explored. Treatment plants could provide the population with water packed in containers.

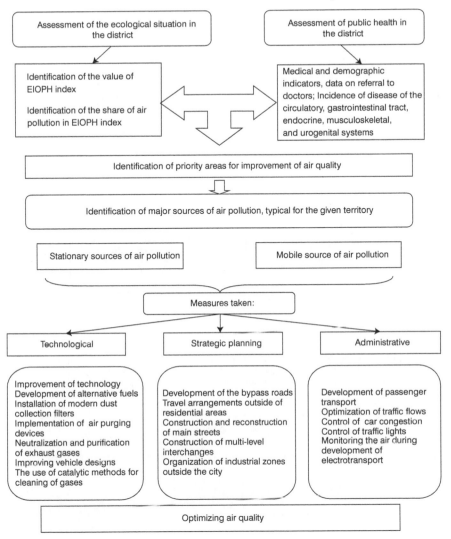

Figure 7.7 Algorithm for improving air quality in priority regions

The practical outcome of the study is to begin the development of optimal plans for providing safe drinking water supplies to Tatarstan and Mari El.

Appendix 1: Methodology for calculating the environmental indicators of public health (EIOPH)

The ranking of territories was conducted according to levels of EIOPH. We determined the significance of EIOPH territories in connection with the methodological recommendations worked out by the F. F. Erisman Federal

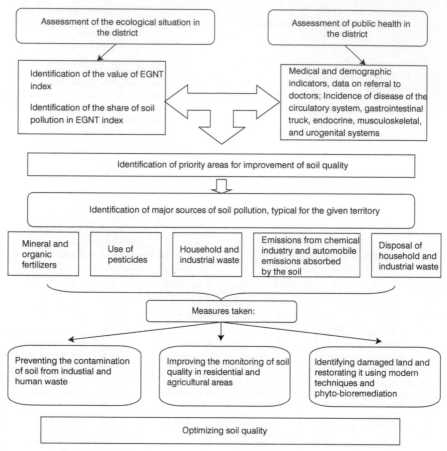

Figure 7.8 Algorithm to optimize the quality of soil by identifying priority areas

Table 7.6 Estimates of potable water requirements in regions of Tatarstan and Mari El

Districts	Population (calculated as a long-term average)*	The required amount of potable water, litres per day**
Republic of Tatarstan		
Agryzskii	36,475	109,425
Atninskii	14,108	42,324
Baltasinskii	33,299	99,897
Bugul'minskii	113,981	341,943
Vysokogorskii	46,255	138,765
Zelenodolskii	160,876	482,628
Mendeleevskii	30,464	91392
Musliumovskii	22,542	67626
Nurlatskii	61,591	184,773
Pestrechinskii	28,172	84,516
Cheremshanskii	21,237	63,711
Chistopolskii	83,327	24,9981
Total	652,327	1,956,981
Republic of Mari El		
Zvenigovskii	44,500	133,500
Novotorial'skii	14,800	44,400
Sernurskii	24,900	74,700
City of Vol'zhsk	56,900	170,700
Total	141,100	423 300

*   According to the materials of the Territorial Departments of the Federal State Statistics Service in Tatarstan and Mari El.
** Calculation of the required amount of drinking water is based on the rate of 3 litres of water per person per day.

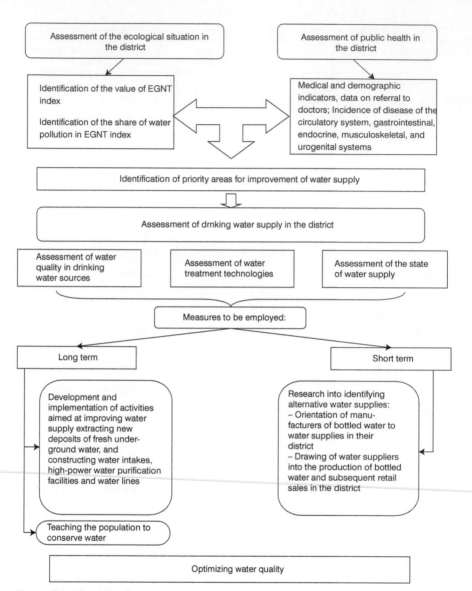

Figure 7.9 Algorithm for optimization of water supply based on the definition of priority areas

Scientific Centre of Hygiene "Evaluation and prognosis of the role of water factors in the formation of public health." This methodological approach was used as a basis for calculating partial scores to determine EIOPH.

Improvements included the expansion of a list of indicators, providing the fullest characteristics of EIOPH in municipal districts.

At the first stage, priority indicators and weights of each factor were identified in the total load. The actual data were grouped by factors such as weather conditions, air, food, water, soil. Grouped data were transferred into dimensionless conditional values (scores) using a three-point scale. The load level for each factor (W_f) was calculated by the formula:

$W_f = a \times k$ where
 a = the degree of disadvantage score;
 k = the weight coefficient of the factor

The resulting conditional values were summed for each factor by the formula:

$W_f = W_{f1} + W_{f2} + W_{f3} + \dots + W_{fn}$

The overall composite index of EIOPH ($W_{average}$) was calculated by the formula:

$W_{average} = W_f / n$, where
$W_{average}$ = the total load on all factors;
 n = the number of considered factors

The proportion of the contribution of drinking water (%) in the EIOPH was calculated by the formula:

$W_{wc} = W_{dw} \times 100 / W_{average}$ where
W_{wc} = the percentage of contribution of drinking water in the EIOPH;
W_{dw} = the load level for drinking water;
$W_{average}$ = the total load on all factors;

The ranking of districts of Tatarstan and Mari El on water quality and air quality and their contribution to the EIOPH was performed using mapping methods taking into account their differences from average regional values.

Appendix 2: Methodology for calculating the relative epidemiological risk in contracting disease

Calculations of the relative epidemiological risk are performed in accordance to the guidelines of the F. F. Erisman Federal Scientific Centre of Hygiene[2] "An assessment of the epidemiological risk to health at the population level in health and hygienic ranking of the territories."

The relative epidemiological risk for each district of the republic was defined as the probability of deviation from the standard rate (background) values. Calculation of the background levels were made on the basis of information on the incidence of the morbidity by classes of internationally classified diseases (ICD) on all the studied districts for five years. The average of three minimal values for each of the pathologies over the last five time intervals was accepted as the background rate.

Values of the relative epidemiological risk were calculated from nonlinear function similar to the normal distribution function:

$$PR_i^J = 1 - \exp\left(-1/2\left(Z^{Ji}_{i+1} / Z^{Fi}_{i+1}\right)\right) \quad \text{where}$$

PR_i^J = relative epidemiological risk of the i-th pathology in area J;

Z^{Ji}_{i+1} = the actual values of the i-th pathology in the area J;

Z^{Fi}_{i+1} = background values of the i-th pathology in the area J;

The size of relative spatial epidemiological risk is obtained by averaging the values for each disease recorded in the entire area of J.

$$PR^J = 1/N \times \sum PR_i^J$$

where

PR^J = the relative risk to territory J with respect to all pathologies;

n = number of pathologies taken into account

Further, depending on the distribution, the entire measurement range of integrated health indices (from 0 to 1) was divided into intervals corresponding admissible (acceptable), moderate, high risk levels. The probability for each value to decline at certain intervals is the same.

Appendix 3: Environmentally and hygienically degraded areas

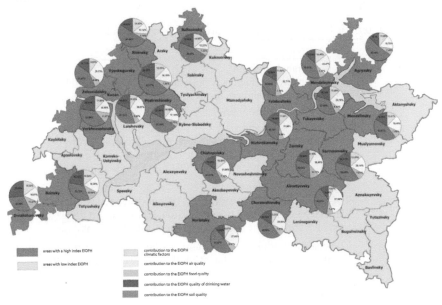

Map 7.1 Environmentally and hygienically degraded areas within Tatarstan's municipal districts

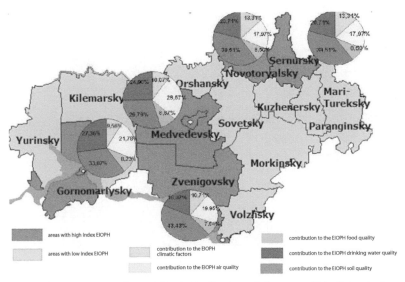

Map 7.2 Environmentally and hygienically degraded areas within Tatarstan's municipal districts and the impact of major pollutants

Appendix 4: Grouping of municipal districts of Tatarstan and Mari El

According to their health ranking in the areas of drinking water and air

Group Name	Characteristics
Group 1	High/favorable air quality Proportion of contribution of the water/air quality in EIOPH was determined as higher than the average national rate and the EIOPH levels were higher than the average for the republic.
Group 2	Low quality water and air In comparison with the average republic indicators, the low quality of drinking water and air, the large contribution of air and water to EIOPH, may lead to levels of EIOPH varying.
Group 3	High-quality drinking water The quality of drinking water/air is higher compared with the average national level; the share of contribution of water in EIOPH is lower than in average in the republic, and the level EIOPH may vary.
Group 4	The quality of drinking water / air is higher, their share of contribution to EIOPH is lower; the value of EIOPH is lower than the regional average.

Appendix 5: Ranking of districts by environmental quality

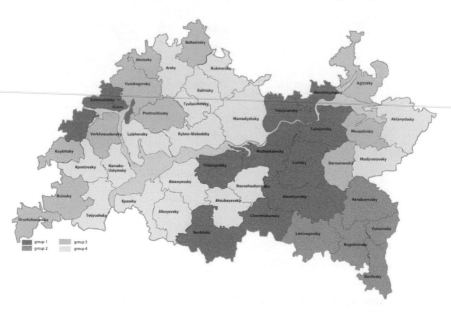

Map 7.3 Ranking of districts in the Republic of Tatarstan by air quality and contribution to hygienic degradation

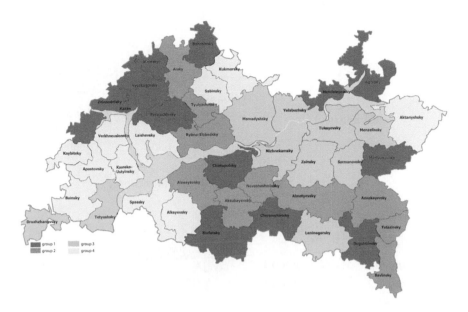

Map 7.4 Ranking of districts in the Republic of Tatarstan by the quality of drinking water and its contribution to hygienic degradation

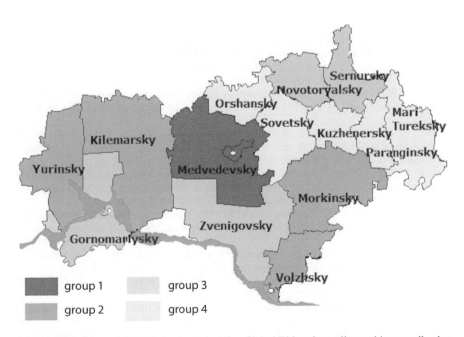

Map 7.5 Ranking of districts in the Republic of Mari El by air quality and its contribution to hygienic degradation

Map 7.6 Ranking of districts in the Republic of Mari El by water quality and its contribution to hygienic degradation

Notes

1 World Health Organization (WHO) International Classifications of Disease: www. who.int/classifications/icd (accessed 1 May 2012).
2 F. F. Dautov, *A Study of the Population's Health in Connection with Environmental Factors,* Kazan': KGU Publishers, 1990, 113.
3 V. P. Rassanov, "A complex evalution of the environment's influence on children's health in Mari El," Candidate of Medical Science dissertation abstract, Kazan, 1993, 18.
4 L. Z. Rashitov, "Hygenic evaluation of the influence of air pollution on the health of pre-school children in the Nizhekamsk and Aznakaev Regions of Tatarstan," Candidate of Medical Science dissertation abstract, Kazan, 2003, 23.
5 L. A. Gabdullina, "Research on the social-environmental situation in urbanized territory with the use of GIS" paper presented at the conference "Ecology: Issues and Paths of Decision," Perm, Russia 28–30 April 2005, 49–51.
6 A. P. Rosolovskii, "Hygienic ranking of territory—the basis of measures for protecting the health of the population," Candidate of Medical Sciences dissertation abstract, Moscow, 2008, 29.
7 N. N. Mitina, "Research status on water resources in the Republic of Tatarstan and environmental security," *Problemy regional'noi ekologii,* 2010, 1, 16–21.
8 Law of the Republic of Tatarstan, 15 January 2009, Environmental Code, No. 5-ZPT http://www.gossov.tatarstan.ru/kodeks/31.03.2011
9 E. A. Bogdan, "Analysis of the role of ecological factors in the development of human capital," *Ekonomika prirodopolzovaniia,* No. 1, 2010, 73-81. See also S. V. Brusnitsyna, "Issues of correlating constitutional law on [maintaining] a favorable environment with constitutional responsibility to preserve nature and the environment," *Zhurnal problemy pravy,* No. 4, 2009, 36–43; M. K. Negobichenko, "The contemporary status of environmental pollution: materials from the interagency scientific council on the ecology of man and the hygiene of the environment in

the Russian Federation," *Ugrozy zdoroviu cheloveka: sovremennye gigienicheskie problem i puti ikh resheniia,* Moscow, 2002, 166–168; K. A. Ballard, "The impact of the environment on health," in *Nursing Administration Quarterly,* 34, 4, Oct-Dec. 2010, 346–350.

10 V. I. Evtushenko, "Environmental migration as the realization of the constitutional right of man and citizen to a healthful environment," *Zakony Rossii: opyt, analiz, praktika,* 6, 2009, 122–124.

11 Dautova, *A Study of the Population's Health.*

12 E. N. Egorova, "Environmental security as a factor of a stable ecological and economic system in the Russian Federation," *Vestnik Kazanskogo gosudarstvennogo finansovo-ekonomicheskoi sistemy,* 3, 2009, 74–77; S. A. Lopatin, "Contemporary problems of megaloposes," *Gigiena i sanitaria,* 4, 2005, pp. 20–25; F. N. Mukhambetov, "Problems of environmental security in the Russian Federation" *IUrist'-Pravoved',* 2, 2010, 121–123.

13 G. G. Onishchenko, "Criteria for [determining] the danger of environmental pollution," *Materialy plenuma mezhvedomstvennogo nauchnogo soveta po ekologii cheloveka i gigiene o.s. RF Ugrozy zdoroviu cheloveka: sovremennye gigienicheskie problemy i puti ikh resheniia,* Moscow, 2002, 5–6.

14 N. A. Matveeva et al., *Gigiena i ekologiia cheloveka,* Moscow: Akademiia, 205, p. 304; E G. Revkova, "Ecological, technogenic and socio-economic risks in the region and their interdependence with the growth of disease among students," *Intellekt, Innovatsii. Investitsii,* 4, 2010, 183–190; E. Bessaloni, F. Vanni, S. Giovannangeli, M. Bessaloni, et al., "Agricultural soils potentially contaminated: risk assessment procedure case studies," *Ann Ist Super Sanita,* 46, 3, 2010, 303–308; J. J. Pignatello et al., "Sources, Interactions, and ecological impacts of organic contaminants in water, soil, and sediment: an introduction to the special series," *Journal of Environmental Quality,* 39, 4, July-August 2010, 1133–1138; R. I. McDonald et al., "Urban growth, climate change, and freshwater availability," *Proceedings of the National Academy of Sciences,* USA, 28 March 2011; D. Valentich, V. Micovich, B. Kolarich, et al., "The role of air quality in perception of health of the local population," *Coll Antropologiia,* 34, April 2010, Suppl. 2:113–117.

15 D. O. Dushkova, "Osnovnye podkhody k provedeniiu mediko-ekologicheskikh issledovanii," *Problemy regional'noi ekologii,* 1, 2010, pp. 246–251; Rosolovskii "Hygienic ranking of territory"; T M. Tikhomirova, "Theoretical and methodological Issues associated with researching the influence of the environment on the population's health," *Ekonomika prirodopol'zovaniia,* 6, 2005, 26–39; D. Nuvolone, et al., "Geographical information system and environmental epidemiology: a cross-sectional spatial analysis of the effects of traffic-related air pollution on respiratory health," *Environmental Health,* Mar. 1; 10–12, 2011.

16 N. N. Beliaeva, "Results and perspectives from research on the histological and cytologic structural-functional indicators of the organism," *Materialy plenuma "Itogi i perspektivy nauchnykh issledovanii po probleme ekologii cheloveka i gigieny o.s.,"* Moscow, 2006, 32–45 and "Cellular status of nose and mouth mucousupon evaluation of various influences on health," *Materialy plenuma "Mezhvedomstvennogo nauchnogo soveta po ekologii cheloveka: sovremmenye gigienicheskie problemy i puti ikh resheniia,"* Moscow, 2002, 34–35; A. Pruss-Ustun, "Knowns and unknowns on the burden of disease due to chemicals: a systematic review, *Environmental Health,* 10, 9, January 21, 2011; P. S. Shah and T. Balkhair, "Air pollution and birth outcomes: a systematic review," *Environment International,* 37, 2, Feb. 2011, 498–516; S. Williams, "Environmental factors affecting elite young athletes," *Medical Sports Science* 56, 2011, 150–170.

17 L. A. Gabdullina, "Research on the social-environmental situation," pp. 49–51; L. P. Stepanova, "An evaluation of the status of the natural environment as a means of identifying zones of environmental degradation," *Vestnik Rossiiskogo universiteta*

druzhby narodov. Seria: *Ekologiia i bezopasnost; zhiznedeiatel'nosti,* 1, 2011, pp. 42–48; V. I. Sturman, "Approaches to expanding the analysis of environmental status using ecological kartography," Moscow, *Geograficheskii fakul'tet,* 2005, 255–263.

18 M. F. Verkhozina, "Indicators of morbidity and mortality as representing the environmental situation in the region," *Problemy regionalnoi ekologii,* 3, 2008, 178–182; M. P. Zakharchenko, "Issues of hygienic diagnostics of environmental status on the basis of dysbiotic occurrences," *Gigiena i sanitaria,* 6, 2004, 50–53; I. A. Ialii, "The application of risk methodologies for controlling the level of environmental security in an urban territory," *Lichnost. Kul'tura. Obshchestvo.* 1, 2009, 310–315; M. Hendrikh, et al. "Pollution sources and mortality rates across rural-urban areas in the United States, *Journal of Rural Health,* Fall; 26, 4, 383–391, 2010; B. Linares et al., "The impact of air pollution on pulmonary function and respiratory symptoms in children," *BMC Pulmonary Medicine,* 2010 Nov 24; 10, 62; C. Washam, "Separating people from pollution: individual and community interventions to mitigate health effects of air pollutants," *Environmental Health Perspectives,* 2011, Jan; 119, 1, A34.

19 A. Cachada, et al. "Sources of potentially toxic elements and organic pollutants in an urban area subjected to an industrial impact," *Environmental Monitory Assessment,* Mar. 2011,17.

20 D. Valentič, et al. "The role of air quality in perceptions of health of the local population," *Coll. Antropol.,* 2010, Apr, 34, suppl 2: 113–117; V. Boldyrev, "Ecological doctrine for the air: what will it be? *Promyshlennye vedomosti: eskpertnaia obshcherossiiskaia gazeta,* 9-10, 2003 and http://www.promved.ru/may_03.html; A. A. Kapustin, "Automobile transport and issues of the environmental situation in large cities," *Mir cheloveka* 1, 2009, 80–94; and M. Mansourian et al. "Air pollution and hospitalization for respiratory diseases among children in Isfahan, Iran," *Ghana Medical Journal,* 2010, Dec; 44, 4, 138–143.

21 L. N. Karlin "Two consensual evaluations of the quality of environmental components in large cities and industrial zones: analysis and comparison," *Uchenye zapiski Rossiiskogo gos. gidrometeorologicheskogo universisteta,* 10, 2009, 67–80; K. Balakrishnan, et al., "Integrated urban-rural frameworks for air pollution and health-related research in India: the way forward," *Environmental Health Perspectives,* 2011 Jan: 119, 1, A12–13; M.R. Gwinn et al., "Meeting Report: Estimating the benefits of reducing hazardous air pollutants: summary of 2009 workshop and future considerations," op cit., 125–130.

22 A. M. Nikanorov, "A complex evaluation of the quality of surface water," *Vodnye resursy* 32, 1, 61–69.

23 N. Alam et al, "Environmental health risk assessment of nickel contamination of drinking water in a rural village in NSW," *NSW Public Health Bulletin,* 2008 Sept-Oct., 19, 9–10, 170–173.

24 T. V. Badaeva, "The quality of drinking water as a risk factor for the health of the population," *Vestnik Rossiiskoi voenno-meditsinskoi akademii,* 2008, 2, 3, 438.

25 G.M. Batrakova, "Methods of assessing social welfare and environmental status of territories where dangerous industriai plants are located," *Nauchnye issledovaniia i innovatsii,* 4, 2010, 15–21.

26 H.Y. Lai, et al., "Health risk-based assessment and management of heavy metals-contaminated soil sites in Taiwan," *International Journal of Environment Res Public Health,* 2010 Oct, 7, 10, 3595–3614.

27 G. M. Batrakova, "Methods of assessing social welfare, 15–21.

28 N. G. Maslennikov (ed.), *Ekologicheskaia toksokologiia,* Moscow, 2004.

29 R. A. Diakaev, "A complex hygienic evaluation of the pollution from heavy metals in the technogenic territories of Bashkortostan," Candidate of Biological Sciences dissertation abstract, Moscow, 2010, 24.

30 L.N. Karlin, "Two Consensual evaluations," 67–80.

31 I. U. A. Rakhmanin, "Pressing problems of complex hygienic characteristics of the city environment and its influence on the health of the population," Materials from the plenum on ecology and hygiene: *"Sovremennye problemy gigieny goroda, medodologiia i puti resheniia,"* Moscow 2006, 14–15.
32 State report on the status of the environment in the Republic of Tatarstan, 2005–2010. : http://eco.tatar.ru/rus/info.php?id=48562 (accessed March 26, 2012).
33 World Health Organization 2010, Report on World Health Statistics, 2010: http://www.who.int/gho/publications/world_health_statistics/EN_WHS10_TOCintro.pdf (accessed 4 May 2012).
34 World Health Organization International Classifications of Disease: www.who.int/classifications/icd/ (accessed 1 May 2012).
35 A. I. Potapov, "Assessment of epidemiological health risk at the population level in the medical-hygienic ranking of territories" in A. I. Potapov et al., *Sbornik metodicheskikh rekomendatsii zdorov'e naselenie i okruzhaiushchaia sreda,* No. 3. Part 1. Moscow: Federal Center of State Sanitary-Epidemiological Monitoring, Ministry of Health of Russia. 1999, 494–503.

Index